SUSTAINABILITY

A History

Jeremy L. Caradonna

OXFORD
UNIVERSITY PRESS

Oxford University Press is a department of the University of Oxford.
It furthers the University's objective of excellence in research, scholarship,
and education by publishing worldwide.

Oxford New York
Auckland Cape Town Dar es Salaam Hong Kong Karachi
Kuala Lumpur Madrid Melbourne Mexico City Nairobi
New Delhi Shanghai Taipei Toronto

With offices in
Argentina Austria Brazil Chile Czech Republic France Greece
Guatemala Hungary Italy Japan Poland Portugal Singapore
South Korea Switzerland Thailand Turkey Ukraine Vietnam

Oxford is a registered trade mark of Oxford University Press
in the UK and certain other countries.

Published in the United States of America by
Oxford University Press
198 Madison Avenue, New York, NY 10016

Library of Congress Cataloging-in-Publication Data
Caradonna, Jeremy L., 1979–
Sustainability : a history / Jeremy L. Caradonna.
p. cm.
Includes bibliographical references and index.
ISBN 978-0-19-937240-9
1. Sustainability—History. I. Title.
GE195.C379 2014
338.9'2709—dc23
2013049859

3 5 7 9 8 6 4 2

Printed in the United States of America on acid-free paper

CONTENTS

ACKNOWLEDGMENTS

Thanks goes, first and foremost, to my wife, Hannah, for her love and tireless support. Our daughters, Stella and Mia, inspired every step of the research and writing of this book. I am hopeful that the world they leave behind one day will be more sustainable than the one they inherited. My parents and in-laws provided much-needed encouragement throughout the writing process. Tim Bent, Keely Latcham, and the entire Oxford University Press staff were a pleasure to work with; their guidance strengthened the manuscript immensely. The manuscript (or portions of it) was also read, critiqued, and improved by Andrew Gow, the participants in the European History Seminar at the University of Alberta, Jake Papineau, Lily Climenhaga, Mike Kennedy, and Emily Kennedy. My sincere thanks for all the useful feedback. Of course, any mistakes in the pages of this book belong entirely to me. I've also benefited from discussions on sustainability with Ulrich Grober, Donald Worster, Chris Turner, David A. Bell, Kurt Rohrig, Paul Hawken, Nathan Perl-Rosenthal, Philippe Lucas, Mary Lucas, David Kahane, Cressida Hayes, Kris Hansen, Sarah Donald (ahem), Chloe Pope Durier, Fabrice Durier, Thomas Shields, Kate Lackey, Jim Fenton,

ACKNOWLEDGMENTS

Ali Cowan, James Cowan, John Nelson, Jody Nelon, Terry Power, Jessie Power, James Gwinnett, Wendy Hoglund, Sarah Blais, Alex Blais, Fiona Williams, James McFarland, Phil Bell, Marcia Bell, Sean Flynn, Eric Hall, Mike Spring. Special thanks to Stella, Mia, Sophie, Angus, Callum, Solomon, Esther, Milo Ignatius, Molly, Ana, Lily, Maya, Oliver, Bea, Sebastian, Hazel, Elliott, Francis, Linnaea, Sebastian, and all the wonderful little kids out there who make me hopeful for the future.

SUSTAINABILITY

Introduction

"We must aim for a continuous, resilient, and sustainable use [of forests]. . . ."

—Hans Carl von Carlowitz, 1713

"Sustainability is a lifestyle designed for permanence."

—Chris Turner, 2010[1]

As hard as it might be to believe, the world once made do without the words "sustainable" and "sustainability." Today they're nearly ubiquitous. At the grocery store we shop for "sustainable foods" that were produced, of course, from "sustainable agriculture"; ministries of natural resources in many parts of the world strive for "sustainable yields" in forestry; the United Nations (UN) has long touted "sustainable development" as a strategy for global stability; and woe be the city dweller who doesn't aim for a "sustainable lifestyle."

Sustainability first emerged as an explicit social, environmental, and economic ideal in the late 1970s and 1980s. By the 1990s, it had become a familiar term in the world of policy wonkery—President Bill Clinton's Council on Sustainable Development, for instance— but the embrace wasn't universal. Bill McKibben, perhaps the most prominent environmentalist of the past 30 years, wrote an opinion piece in the *New York Times* in 1996 in which he dismissed sustainability as a "buzzless buzzword" that was "born partly in an effort to obfuscate" and which would never catch on in mainstream society. In McKibben's view, sustainability "never made the leap to lingo"— and never would. "It's time to figure out why, and then figure out something else." (McKibben preferred the term "maturity.")[2] Many others have since accused "sustainability" and "sustainable development" of being superficial terms that mask ongoing environmental degradation and facilitate business-as-usual economic growth. Those are debatable points that will be discussed in this book. But one thing is clear: McKibben was quite wrong about the quick decline of "sustainability."

One way to demonstrate this growing interest is to look at book titles that bear the word "sustainable" or "sustainability." It's difficult to find books published before 1976 that employ these words as titles or even as keywords.[3] Indeed, as Figure 1 shows, no book in the English language used either term in the title before 1970. But since 1980 there has been an explosion of books and articles that not only use those words as titles but also deal with the many facets of sustainability. Indeed, thousands of books make up this growing body of literature. What's more, a quick Google search for "sustainability" returns around 150 million hits.

Is sustainability a buzzword? Absolutely. But is it also "buzzless"? Assuredly not. Governments, communities, organizations, and individuals all over the world have sought to align themselves with the basic principles of what they call "sustainability"—a desire to create a society that is safe, stable, prosperous, and ecologically minded. The practices inspired by the concept of sustainability could give

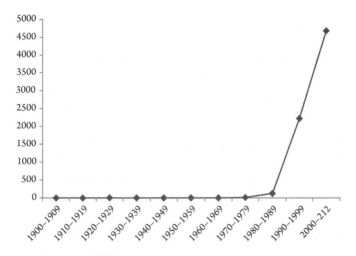

Figure 1. Books with the word "sustainable" or "sustainability" in the title, 1900–2012.

rise to the world's third major socio-economic transformation, after the Agricultural Revolution that took place 10,000 years ago, and the Industrial Revolution(s) of the late eighteenth and nineteenth centuries. It is not only a buzzword but also a galvanizingly powerful term whose application subsumes a number of other movements, environmental perhaps most of all.

"Sustainability" is, first and foremost, used as a corrective, a counterbalance, and directly tied to climate change (a term that most governments would like to avoid). Those who use it argue that we are 250 years into an "unsustainable" ecological assault on the planet that was triggered by industrialization and that has left us with a lot of soul searching and cleaning up to do. "Sustainability" therefore is a way of acknowledging how humankind has created an imbalance. According to Jeffrey D. Sachs, we now live in the Age of the Anthropocene, in which "human activity" has become the "dominant driver of the natural environment."[4] We *are* or have become a kind of natural disaster.

The Fifth Assessment Report (2014) from the Intergovernmental Panel on Climate Change, a team of scientists whose job it is to sort through and summarize the state of climate science, makes it clear that Earth's climate system is warming steadily due to "anthropogenic greenhouse gas concentrations," such as carbon dioxide, methane, and nitrous oxide, all of which trap heat (at infrared wavelengths) that would otherwise escape from the Earth's atmosphere. "It is extremely likely that human influence has been the dominant cause of the observed warming since the mid-20th century," the report concludes.[5] Climate change has already begun to alter natural systems and the environment in troubling ways: increasingly unpredictable temperatures and weather patterns, changes in the hydrological cycle that generate droughts and larger and more frequent storms, rising sea levels from melting ice caps, the die-off of some species, and so on. Climate change also suffers from "positive feedback cycles" (the loss of reflective ice cover, the release of natural stores of methane, etc.) that act as a domino effect and accelerate the speed of global warming. If greenhouse gas emissions continue to grow, average global temperatures could rise by 8.1° F by 2100.[6] Furthermore, the growing population of homo sapiens on the planet, which surpassed the 7 billion mark in 2012, combined with man-made pollutants and the appropriation of over 30% of the net primary production of organic material—that is, we use or alter much of what nature has to offer—has resulted in devastating consequences for the world's ecosystems.[7]

This is what sustainability is meant to counteract: a moribund economic system that has drained the world of many of its finite resources, including fresh water and crude oil, generated a meltdown in global financial systems, exacerbated social inequality in many parts of the world, and driven human civilization to the brink of catastrophe by unwisely advocating for economic growth at the expense of resources and essential ecosystem services.[8]

Those whom we'll call "sustainists"—from scientists and engineers to economists, educators, policymakers, and social activists—have taken on the many challenges listed above. What they seek—and how—is always a subject of intense debate, even among themselves, but the broad contours are easy to articulate: safe and livable cities with abundant green spaces; buildings that produce their own energy; public transportation networks to decrease reliance on cars; agricultural systems that can produce enough food to meet human needs without genetically modified organisms or monoculture and without degrading soils and waterways with petrochemicals; and a healthy environment. To sustainists, sustainability means planning for the future and rejecting that which threatens the lives and well-being of future generations. It means creating a "green," "low-carbon," and "resilient" economy that runs on renewable energy and does not support growth that would impair the ability for humans and other organisms to live in perpetuity on the Earth.[9] For many it has a utopic dimension: decentralized forms of democracy that support peace and social justice.[10]

In short, for those who embrace sustainability in the fullest sense—as an environmental, social, economic, and political ideal—we're at a crossroads in our civilization. There are two paths to take: continue with business as usual, ignore the science of climate change, and pretend that our economic system isn't on life support or remake and redefine our society along the lines of sustainability.

This is a book about the making of the sustainability movement, and to cover what is involved, we have to use the term "movement" in the broadest sense of the word. Protesters marching and holding signs or occupying public spaces are only part of it. Rather, it encompasses the development and application of the concept of sustainability in a broad range of domains: urbanism, agriculture and ecological design, forestry, fisheries, economics, trade, population, housing and architecture, transportation, business, education, social justice, and so on. This book considers how sustainability went from being a relatively marginal idea to being the centerpiece of international accords; a top priority for governments, corporations, and nonprofit organizations; and a philosophy of hope and resilience with widespread appeal.

This book will give a historical account of the growth of this movement: where it came from and how it took shape. While it is of rather recent origin, the ideas that undergird it developed over a long period of time. The UN conferences and commissions that have put sustainability on the agenda of the international community in the past 30 years were, in a sense, the result of three centuries of debate about the relationship between humanity and the natural world. We cannot understand the contemporary sustainability movement without first understanding the historical events that made sustainability thinkable.

The conceptual roots of sustainability stretch back at least to the late seventeenth century. One of the main goals of this book is to uncover the intellectual developments that have shaped the movement.[11]

We should not assume that sustainability was a necessary outcome or that industrial society was destined to embrace this idea to the degree that it has, but the growing importance of the "sustainability revolution" is tied to its historical development.[12] Most studies of the concept of sustainability, by contrast, dedicate less than a paragraph to discussing its past, and many writers seem to assume that the idea appeared, ex nihilo, for the first time in 1987, when Gro Harlem Brundtland and the UN-backed World Commission on Environment and Development released a hugely influential document called *Our Common Future*, which offered the first well-developed definition of "sustainable development."

Yet the definition of sustainability has been a subject of intense debate ever since the late 1980s. Is it an end point or a process?

What is considered sustainable versus unsustainable? Who gets to make these determinations? It has become a commonplace in the literature on the subject to suggest that the definition is too vague and thus susceptible to exploitation and "greenwashing."[13] It is certainly true that sustainability is a broadly conceived philosophy. In this sense, it is a bit like "democracy," "justice," or "community," all of which are discursive fields that suggest a set of conditions rather than a specific outcome. As we will see, in the marketplace of ideas, breadth has been advantageous for sustainability.

A helpful place to begin is with etymology. Both "sustainable" and "sustainability" derive from the Latin *sustinēre*, which combines the words *sub* (up from below) and *tenēre* (to hold), and means to "maintain," "sustain," "support," "endure," or, perhaps most poignantly, "to restrain." From Latin, the word passed to Old French as *sostenir* and then to modern French as *soutenir*. (Similar linguistic developments occurred in other Romance languages in the Middle Ages.)[14] From French, the word passed to English as the verb "to sustain" and was in widespread usage by the Early Modern period; it can be found in John Evelyn's influential treatise on forestry called *Sylva* (1664), for instance. The *Oxford English Dictionary* states that the adjective "sustainable" entered common usage in 1965 via an economics dictionary that used the phrase "sustainable growth." The noun "sustainability" entered English in the early 1970s. The coining of these neologisms is an important indication that this verb ("to sustain") had developed by the latter part of the twentieth century into an identifiable concept (maintaining human society over the long term). It is also worth noting the parallel etymology of the word in German: *nachhaltig* (sustainable) and *Nachhaltigkeit* (sustainability) both entered the Saxon dialect of German in the eighteenth century via Hans Carl von Carlowitz's works on sustained yield forestry.

7

Nearly all of the definitions of sustainability that have circulated in recent years emphasize an ecological point of view—the notion that human society and economy are intimately connected to the natural environment. Humans must live harmoniously with the natural world if they—or we—hope to persist, adapt, and thrive indefinitely on the Earth. Rather than viewing society and the environment as separate or even antagonistic spheres, the concept of sustainability assumes that humans and their economic systems are indelibly linked. The most common model of sustainability to emerge in recent years is a tripartite Venn diagram that illustrates the interconnectedness of the "three Es": environment, economy, and equity or social equality (see Figure 2). This model was endorsed by the 2005 UN World Summit and appears in countless books, websites, and ecological models. Sometimes a fourth "E," education, is added to the diagram to reflect the importance of education in establishing a sustainable society.

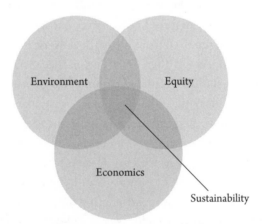

Figure 2. The three Es of sustainability represented in a diagram. A "sustainable" society requires a balance between and equal concern for the environment, social equality, and the economy.

A newer model reconceptualizes the diagram as a series of concentric circles, in which the environment is seen as the foundation of sustainability, with society and the economy nested inside (see Figure 3). The latter model reflects the critique by sustainability economists, such as Peter Victor and Herman Daly, who argue that society and the economy are supported by and could not exist without the environment, and therefore that the environment should take conceptual priority in any model of sustainability. As Daly puts it, "All economic systems are subsystems within the big biophysical system of ecological interdependence."[15]

A number of economists, ecologists, scientists, and organizations have offered more precise definitions of sustainability.[16] In 1989, the Swedish oncologist Karl-Henrik Robèrt founded a highly influential organization called the Natural Step that is dedicated to promoting a sustainable society. Robèrt and his colleagues have outlined "four systems conditions for sustainability" that guide their consulting and advocacy efforts around the world.

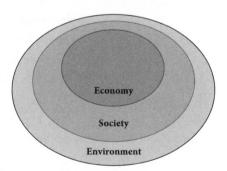

Figure 3. This diagram places the environment at the foundation of the model. It emphasizes that human society and the economy cannot exist without the environment, and therefore it takes conceptual priority.

In a sustainable society, nature is not subject to systematically increasing (1) concentrations of substances extracted from the earth's crust (digging), (2) concentrations of substances produced by society (dumping), (3) degradation by physical means (destroying), and (4) people are not subject to conditions that systematically undermine their capacity to meet their needs.[17]

The Natural Step uses a scientific framework to help uphold the integrity of ecosystems and is geared toward shaping the practical decisions of businesses, organizations, and governments. As such, it says rather little about the role of individuals in creating a sustainable society, which stands in contrast to some of the definitions put forth by others.

Richard Heinberg, perhaps the world's leading expert on peak oil and a senior fellow-in-residence at the Post Carbon Institute, has put together five axioms ("self-evident truths") of sustainability:

(1) Any society that continues to use critical resources unsustainably will collapse. (2) Population growth and/or growth in the rates of consumption of resources cannot be sustained. (3) To be sustainable, the use of renewable resources must proceed at a rate that is less than or equal to the rate of natural replenishment. (4) To be sustainable, the use of nonrenewable resources must proceed at a rate that is declining, and the rate of decline must be greater than or equal to the rate of depletion. (5) Sustainability requires that substances introduced into the environment from human activities be minimized and rendered harmless to biosphere functions.[18]

One can see here the obvious affinities to the conditions laid out by the Natural Step. Both definitions emphasize the need for society to conserve resources, protect ecosystems, and minimize pollution, although Heinberg's definition lacks the element of social equity that the Natural Step includes in its fourth condition.[19]

The physicist Albert A. Bartlett has developed perhaps the most elaborate definition of sustainability, which involves several laws, hypotheses, observations, and predictions and appears in an essay from 1997–1998 called "Reflections on Sustainability, Population Growth, and the Environment—Revisited." His definition is far too complex to summarize neatly but it focuses on the risks that unchecked population growth, economic growth, and fossil fuels pose to long-term human existence on the planet. Bartlett is best known for arguing that the term "sustainable growth" is an oxymoron—a belief shared by many sustainability economists—and for his sardonic contention that "modern agriculture is the use of land to convert petroleum into food." He concludes his discussion by reiterating the need to limit population growth, to "make [economic] growth pay for itself" and to "improve social justice and equity." What's noteworthy about Bartlett's definition is that it focuses less on "the environment" and more on economic growth, population, agriculture, and energy principles.[20]

Finally, there is John Dryzek's interpretation of sustainability. In *Politics of the Earth*, Dryzek argues that there are several competing discourses on the environment in current circulation. Dryzek borrows the term "discourse" from the French philosopher Michel Foucault, who used the word not in its ordinary meaning of "dialogue" or "debate" but to signify a way of talking about a body of knowledge, one that takes shape over time, generates categories and terminology, and has, at least in theory, a very formative impact on a culture's (or an individual's) sense of what is true, real, and essential.[21] Dryzek argues that there has been a shift in environmental discourses over the past few decades, away from a focus on wilderness, preservation, and population growth and toward energy supply, animal rights, species extinction, anthropogenic climate change, depletion of the ozone layer, toxic waste, the protection of whole ecosystems, environmental justice, food safety, and genetically modified organisms. In the course of this shift, several relatively new discourses have taken shape, alongside an older Prometheanism (the idea that natural resources are unlimited and markets can solve all environmental problems): green radicalism, survivalism, problem solving, and sustainability. Dryzek defines the latter as an "imaginative and reformist" discourse that attempts to eliminate the conflict between economic and environmental values. He also demonstrates the pluralism of sustainability and the deep-seated disagreements—on such topics as economic growth—that take place within it.[22]

Dryzek thinks of sustainability as a broad debate rather than a specific model, system, or idea. Nonetheless, there are a number of common terms, categories, and principles that recur in discussions about sustainability. The four main features or principles of these discussions are set out below and form the intellectual foundation of the sustainability movement. Identifying these four features is therefore key to understanding what this book historicizes.

HUMAN SOCIETY, THE ECONOMY, AND THE NATURAL ENVIRONMENT ARE INTERCONNECTED

This is the essential idea in the diagrams on the three Es shown above. It has roots in the science of ecology, which, as Donald Worster has shown in *Nature's Economy: A History of Ecological Ideas*, stretches back to the eighteenth century, even before the word "oecologie" was coined in 1866.[23] Ideas about "nature's economy" have passed through several stages of intellectual development since the 1700s. The idea of an ecosystem, in which living organisms and nonliving components are tied together by nutrient cycles and energy flows, is a relatively new idea in ecology and one that has had a profound impact on the "systems thinking" of sustainability. The three Es is essentially an ecological idea that stresses the dynamic interaction between human communities, the flow of resources, and the natural environment.

Sustainability involves more than "the environment"; it is equally interested in social sustainability (often summarized as well-being, equality, democracy, and justice) and sensible economics but, above all, in the interconnectedness of these domains.[24] Indeed, the field of sustainable development has generated overlapping definitions of economic, environmental, and social sustainability. The economic dimension requires, for instance, a system that can produce goods and services on a continuous basis, avoid excessive debt, and balance the demands of the different sectors of the economy. The environmental dimension requires the maintenance of a stable resource base, the preservation of renewable resources and the "sinks" that process pollution, and the safeguarding of biodiversity and essential ecosystem services. Finally, the social dimension of sustainability involves a range of factors, including a fair distribution of resources, equal opportunities for all citizens, social justice, health, mental well-being and the ability to live a safe and meaningful life, access to education, gender equality, democratic institutions, good governance, and political participation.[25] In short, for a society to be considered sustainable, it must address not only environmental but also social and economic issues.

A SOCIETY WILL RESPECT ECOLOGICAL LIMITS OR FACE COLLAPSE

The idea of limits is a direct response to the assumption in classical economics and industrialism that nature is essentially a cornucopia, that natural resources can never run out (or that market prices and technology will always "save us"), that overconsumption is not a problem, and that the human population can continue to grow indefinitely. As economist Julian Simon put it in 1997, "The material conditions in life will continue to get better for most people, in most countries, most of the time, indefinitely."[26] Economists and ecologists began to question economic and population growth in the mid-twentieth century, but the work that is most closely associated with these concerns is the Club of Rome's 1972 bombshell, *The Limits to Growth*. In that book, systems theorists Donella Meadows, Dennis Meadows, Jørgen Randers, and William W. Behrens III argued that the world's growth-obsessed society was hitting a wall. "If the present trends in world population, industrialization, pollution, food production, and resource depletion continue unchanged, the limits to growth on this planet will be reached sometime within the next 100 years. The most probable result will be rather sudden and uncontrolled decline in both population and industrial capacity," they wrote. Jared Diamond's *Collapse* reminds us what happens to societies that live beyond their means.[27]

The iconoclastic work of the Club of Rome, combined with the writings of other ecological economists in the late 1960s and 1970s, challenged conventional economic thinking and forced a global debate on the drawbacks of growth.[28] The basic idea that humans need to live within limits is now a basic assumption of sustainists, even though divisions on the question of economic growth remain marked.[29] In the 1990s, Daly laid out three simple rules that define the limits to energy and material throughput that are now common fare in the literature on sustainability:

- For a renewable resource—soil, forest, fish—the sustainable rate of use can be no greater than the rate of regeneration of its source.

- For a nonrenewable resource—fossil fuel, high-grade mineral ores, fossil groundwater—the sustainable rate of use can be no greater than the rate at which a renewable resource, used sustainably, can be substituted for it.
- For a pollutant the sustainable rate of emission can be no greater than the rate at which that pollutant can be recycled, absorbed, or rendered harmless in its sink.[30]

A SOCIETY THAT HOPES TO STICK AROUND LONG TERM NEEDS TO PLAN WISELY FOR THE FUTURE

The intergenerational aspect of the sustainability movement takes its inspiration, in part, from the Iroquois Confederacy's thousand-year-old oral constitution that requires chiefs to consider the impact of their decisions on distant future generations: "In every deliberation, we must consider the impact on the seventh generation." The idea that a society should plan for the future—that it should *not* "mortgage its future" or create undue burdens on future humans—is part of the ethical consciousness of sustainability. Sustainability advocates argue that actions likely to create social, economic, and environmental harm—unchecked deforestation, the creation of radioactive waste, emitting large quantities of ozone-depleting chlorofluorocarbons and greenhouse gases, and so on—are unethical because they force upon future generations (in addition to our own) problems that would not have otherwise existed. It is unethical to benefit at the expense of our yet-to-be-born descendants. *Our Common Future* famously used intergenerational language in its definition of "sustainable development": "Humanity has the ability to make development sustainable to ensure that it meets the needs of the present without compromising the ability of future generations to meet their own needs."[31] The industrial ecologist John Ehrenfeld argues that we need to assume the "possibility that human and other life will flourish on the planet forever."[32]

A literary variant of this idea appears in Alan Weisman's *The World Without Us*, which suggests that human society should be considered sustainable only if, in the event that all humans were suddenly to disappear from the Earth, the remnants of society wouldn't impede the ability of other species (flora and fauna) to flourish.[33] As it stands now, a mass die-off of homo sapiens would have devastating effects on global ecosystems. Think of, for instance, the more than 440 nuclear power plants worldwide that would either explode or melt down. But, more prosaically, sustainability demands more prudence, more foresight, less entitlement, and less selfishness in socio-political, economic, and environmental planning.

LOCALIZE, DECENTRALIZE

Sustainability, as an idea and a movement, is a reaction against the perceived unsustainability of industrial society (or at least many of its core features). The proposed idea of returning to small-scale energy production, local agriculture, decentralized decision making, and low-impact practices is not directed at, say, the highlanders of New Guinea, who have lived in harmony with their surroundings for over 40,000 years.[34] The idea that "small is beautiful" is a radical notion only in the context of a massive industrial society that concentrates power in the hands of elites, centralizes and transports resources over long distances, and operates with the assumption that energy production must come primarily from dirty and nonrenewable fossil fuels. Similarly, the "buy local" movement is a reaction to the growing dominance of international agro-business and industrial conglomerates. The idea of going small and local—it's almost always conceptualized as a "return" of sorts—stems from an awareness that the conventional practices of industrialism cannot continue in their present form forever.[35] Thus reemphasizing the small and the local is a way of rejecting one of the core assumptions of the Industrial Revolution and modernity: the idea that a large and centralized society running on brown energy is basically an unstoppable juggernaut. By contrast, sustainists see industrial society as weak and vulnerable to collapse, while the reorientation toward the local is offered as a strategy for societal resilience.

The concept of sustainability is here heavily influenced by E. F. Schumacher's *Small Is Beautiful: Economics as if People Mattered*, which made a huge splash in political and economic circles when it was first published in 1973. Schumacher, who was the chief economic advisor to the United Kingdom's National Coal Board, made an about-face against classical economics and attacked the developed world's centralized energy production, overreliance on fossil fuels, public disempowerment, and the fanatical faith in unlimited economic growth. "The substance of man cannot be measured by Gross National Product," he wrote in one famous passage. Elsewhere he noted:

> The economics of giantism and automation is a left-over of nineteenth-century conditions and nineteenth-century thinking and it is totally incapable of solving any of the real problems of today. An entirely new system of thought is needed, a system based on attention to people, and not primarily to goods— (the goods will look after themselves!). It could be summed up in the phrase, "production by the masses, rather than mass production."

Schumacher took issue with all forms of centralized "giantism"—
totalitarianism, command economies, and oligarchic capitalism—
and offered a framework for a new economic system that focused on
small-scale political units and technologies, local decision making
and energy production, self-sufficiency, and the humanization of
work. "It is moreover obvious that men organized in small units
will take better care of their bit of land or other natural resources
than anonymous companies or megalo-maniac governments which
pretend to themselves that the whole universe is their legitimate
quarry."

Schumacher's philosophy of economics had a huge impact on
the sustainability movement. But even beyond the economic considerations, a return to the small and the local defines an ethos
of duty and empowerment. As Dryzek reminds us, sustainability
assumes action on the part of individuals—a decentralization of
responsibility—rather than a passive attitude that expects governments or "someone else" to solve our problems. The idea of "powering down"[36] our society—of simplifying, of reappreciating local
culture and resources—has a multiplicity of applications that
range from food consumption (e.g., the 100-Mile Diet) and energy
creation (e.g., Net-Zero homes) to localized decision making
processes (e.g., municipal sustainability action plans and neigh-
borhood land use committees) and support for local and small
businesses.

The small and the local have been applied and interpreted in a
number of ways over the past few decades. Consider the ninth of
Holmgren's permaculture design principles: "Use small and slow
solutions." Rifkin touts the benefits of a "distributed low-carbon
era" and "lateral power" that is already replacing "the traditional, hierarchical organization of economic and political power."[37] Lester
R. Brown advocates for "greater local self-reliance."[38] This is not to
say that the big and the centralized could not have their place in a

sustainable society; rather, it is a repudiation of the idea that society must always have the centralized and large-scale technologies as default modus operandi.

The sustainability movement generally functions with all four of these assumptions in mind, although, of course, there is broad debate about the specifics. Indeed, there is a surprising range of viewpoints and conflicts that occur within the umbrella concept of sustainability, as even the small sampling of literature and theories discussed above suggests. This book will examine the contours and complexities of how these concepts evolved and the origin of and relationship between historical ideas and the current sustainability movement.

I offer two final points. First, sustainability is not just another term for environmentalism, nor is the history of sustainability the same thing as environmental history.[39] Sustainists are trained to look at complex systems and find relationships between society, economy, and the natural world. Sustainability and environmentalism share a common history to a certain point, but the sources of sustainability go well beyond the canon of thinkers who have shaped the environmental movement (John Muir, Aldo Leopold, Rachel Carson, Barry Commoner, etc.). The history of sustainability is as much social, political, and economic as it is environmental history.

Second, though the idea of sustainability has taken shape in many parts of the world and the sustainability movement is today fully globalized, this book deals primarily with Europe and North America. This is not quite a global history. Non-"Western" societies such as the Highlanders of New Guinea, the Iroquois Confederacy, China, and Japan appear, but the focus is on the way in which sustainability emerged as a constructive reaction to unsustainable

European and colonial industrial society.[40] As noted above, sustainability presupposes an industrial present that cannot endure—the realization that current approaches will not hold up over time. It is essentially a response to perceived deficiencies within modernity and industrialism—"progress" defined as consumption, the population explosion, environmental degradation, economic growth at the expense of ecosystems, the extinction of species, extreme social inequality, unstable economic systems, pollution, a throwaway society, and so on. An inherently sustainable society would not need an explicit movement. This is a book for and about societies that are looking to restore balance and create stability. Industrial societies can never go back to some idealized, pre-industrial ecotopia. But through studying the history and development of the sustainability movement, we can chart a path to the sustainable future.

Chapter 1

Loath This Growth:
Sources of Sustainability
in the Early Modern World

This chapter begins in the period that historians of Europe and the Atlantic world call "early modernity" (seventeenth and eighteenth centuries). It could have begun in the Middle Ages, with the hunting reserves and protected forests established by European rulers in Venice and elsewhere. It could start with an analysis of indigenous societies, from Easter Island to the Maya, that failed to live sustainably and eventually collapsed. It could even begin in antiquity, with Pliny the Elder and his encyclopedic *Natural History* that tells us so much about Roman conceptions of the natural world.

But we begin in the early modern period because of the clear linkages between the modern sustainability movement of the twenty-first century and the consciousness and practices that developed in early modernity. After all, the concept of "sustainability" was given a name in the early eighteenth century by a Saxon bureaucrat who coined the term "Nachhaltigkeit" to describe the practice of harvesting timber continuously from the same forest. Indeed, sustained yield forestry took shape at this time not only in Western Europe but also in Japan, around other parts of Asia, and on colonial islands in both the West and East Indies. The practice of exploiting forests sustainably was but one indication of an incipient awareness about the value of living within biophysical limits and the need to counteract resource overconsumption. Many documents that survive from this period demonstrate that it was possible to have at least a rudimentary idea about the complex relationship between social well-being, the economy, and the natural world. That is, the "systems thinking" of sustainability—the method of studying complex, interrelated systems—clearly has roots that stretch back to this largely pre-industrialized world.

In 1700, the global population of homo sapiens was somewhere between 600 million and 650 million. Beijing might have approached a population of 1 million, which would have constituted a megacity at the time, but most "cities" had fewer than 50,000 inhabitants. Paris had been the largest city in Europe for some time, but London, with its 575,000 souls, surpassed Paris around this time and continued to swell. Still, most people on the planet lived either in rural areas or in small settlements, even in fairly urbanized areas such as Europe. Countless peasants would never glimpse a city. There was no electricity, no telephones, no internal combustion engines, no synthetic polymers, no fossil-fueled flying machines, and no mass media, although there were a few newspapers and journals in Europe and elsewhere.

Certainly, the pastoralist, hunter-gatherer, and agricultural societies that dotted the globe all impacted the environments in which they lived and worked. But in a world before industrial production, plastic, nuclear waste, and synthetic chemicals, pollution rates and environmental degradation were, on the whole, considerably less severe than they are today.[1] Based on the pioneering work of William Cronon and other environmental historians, though, we know that even pre-industrial indigenous societies altered the landscapes that sustained them. Nature was not in a state of static perfection before the arrival of European axes and industry. As Cronon writes, "There has been no timeless wilderness in a state of perfect change-lessness."[2] Aboriginal societies in the Americas used strategic deforestation for hunting (and agriculture) and practiced widespread agroforestry.[3] Even pre-industrial, medieval Europe faced the growing problem of woodland loss. England was overwhelmingly deforested by the thirteenth century, and in fact deforestation was already a problem in many ancient societies, including Greece and the Roman Republic.[4] One recent estimate places woodland at only 16% of the total land in eighteenth-century France, a country that had once been covered in dense forests.[5] Most of the world, especially the Northern Hemisphere, had to cope with the effects of the Little Ice Age that lasted from the fourteenth century to the early nineteenth century, which created erratic temperatures and affected agricultural production.[6]

All of this is to say that the world before the Industrial Revolution had its ecological problems, too, even if they paled in comparison to the crisis faced by the planet today. Many societies before the nineteenth century dealt with deforestation, desertification, soil erosion, silted rivers, urban air pollution, drought, and intermittent crop failure. As Jared Diamond demonstrates in his best-selling book, *Collapse: How Societies Choose to Fail or Succeed*, a whole series of global societies collapsed as a result, in part, of overstressing local environments. Diamond formulates a five-point framework to understand the collapse of such historical societies as those living on Easter Island, Pitcairn Island, and Henderson Island (all located in the South Pacific), the Anasazi Native Americans who lived in present-day New Mexico, the Maya civilization of the Yucatán and surrounding areas, and the Vikings who populated southern Greenland. The five factors are as follows: environmental damage, climate change (man-made or non-man-made), hostile neighbors, friendly trade partners (or lack thereof), and the society's response to its environmental problems.[7] The final point is an important one because reacting to a problem requires that a society or part of it first acknowledge its own faults. A certain consciousness needs to develop before anything can be done to get off the path of unsustainability.

It is the contention of this chapter that at least some early modern pre-industrialized (or barely industrialized) societies—or some elements within those societies—recognized the patterns of what we would today call "unsustainable living" and began to react to those patterns with constructive criticism and innovative practices. What's so striking about the eighteenth century, especially in Europe, is that it witnessed the genesis of an unsustainable, growth-based, industrialized society *as well as* a powerful set of practices and counter-discourses that illuminated an alternative path. To be clear, this is *not* to imply that England, France, or the Germanic states were (or are) sustainable societies simply because they began to recognize and respond to some unsustainable practices. Although none of those countries have collapsed in the way that Norse settlements did in Greenland, they also shouldn't be romanticized as ecotopias. For instance, the French and English, after deforesting and exhausting their own lands, simply seized and exploited untapped resources in the colonial world.[8] These complexities notwithstanding, changing attitudes in early modernity toward humans and their relationship to the natural world have served as intellectual sources of the modern sustainability movement, even if they did not succeed in placing "advanced" economies on the path to ecotopia.

In a sense, then, we have inherited from early modernity two different yet sinuously interconnected cultural legacies. On the one hand, the eighteenth century was the period that witnessed the birth of the Industrial Revolution, modern growth-based economics, and what has recently been termed the "consumer revolution." Although new machines and techniques had a limited impact on the economy before the nineteenth century, modern manufacturing came into being in the century before 1800, as steam engines and other machines began to supplant animal labor and as factory-based wage labor gradually replaced the long-tenured artisanal workshop system. In England and America, inventors produced new machines that would eventually revolutionize the global economy and change the course of human history. The spinning jenny (1764), the steam engine (1769), the water frame (1771), the power loom (1785), the cotton gin (1794), and other inventions paved the way to modern forms of transportation and industrial production. In economics, the classical theory of capitalism took shape in the pages of Adam Smith's 1776 *An Inquiry into the Nature and Causes of the Wealth of Nations* and in the writings (and governmental policies) of the French physiocrats, such as Anne-Robert-Jacques Turgot, who defended the idea of free markets, economic growth, and a strict division of labor that spurred productivity.

In terms of consumption, British historians have narrowed in on the period after 1690 as the beginning of a "consumer revolution," in which Europeans grew rich off of colonial trade and began to consume culture and products (coffee, tobacco, sugar, textiles, fine goods) on a mass scale.[9] Of course, much of this colonial wealth and economic growth was rooted in slave labor, and perhaps 50% of all the slaves taken from Africa in the early modern period left in the eighteenth century, transported across oceans on French, British, and Portuguese ships.[10] Colonial capitalism also put on display the animosity (or at least apathy) toward the natural world felt by many Europeans, especially on tropical islands where forests were wantonly "subdued" and cleared for the sake of planting valuable cash crops. It was all part of the process called "ecological imperialism" by Alfred Crosby.[11]

On the other hand, the eighteenth century is the source of many of the embryonic ideas that today inform the sustainability movement. This is the period in which abolitionism formed as a powerful critique of slavery and the concept of human rights circulated widely in the Atlantic world, taking center stage in what are today called the Atlantic revolutions: the American Revolution, the French Revolution, the Haitian Revolution, and significant uprisings in the Austrian Netherlands (Belgium), the Netherlands, Ireland, and elsewhere.[12] Where would the concept of social justice be without abolitionism and these democratic revolutions? Moreover, while the eighteenth century is remembered as the period in which classical economics took shape, there were nonetheless critics of economic liberalism in the later part of the century who were able to correlate changing economic policies to social and "environmental" problems. To be sure, a critique of economic growth did *not* appear out of thin air in the late twentieth century (see chapter 4). Furthermore, not everyone endorsed the new consumer society, and many moralists in Europe took aim at the vanity and decadence that came with materialism, greed, and consumption. There were even those who connected greed and overconsumption to deforestation. Indeed, this is the period in which forestry became a legitimate science, woodland overconsumption became a widely recognized problem, and sustained yield forestry became an official policy of governments in many parts of the world. This is also the period in which views of the natural world underwent a conceptual revolution, at least in the Western world, as religious conceptions of a "created" Earth gave way to a more secular Enlightenment perspective that viewed the natural world as inert and in need of domination but ultimately useful, governed by natural law, and knowable in a systematic way. Finally, this is the period in which some critics of urbanism and scientific progress, such as the Swiss philosopher Jean-Jacques Rousseau, took to the woods and began to craft a kind of "natural religion" that valorized the natural world and simple living.

These two legacies of early modernity are neither simplistic nor even mutually exclusive. The period that we think of today as the Enlightenment is fraught with contradiction, and it isn't so easy to determine what one might consider the "good" versus the "bad" elements of this complex heritage. To be clear, there was no explicit sustainability movement (or even environmental movement) in the eighteenth-century Western world. Nor was there a holistic conception of ecology, as there was in many of the indigenous societies that were in the process of being brutalized by European imperialists. Moreover, the people and documents studied in this chapter exhibited an anthropocentrism typical of the period; few if any Europeans wrote of the "rights of nature" or even cared about nature on its own terms, exterior to the needs of humans, who were still conceptualized as separate from and dominant over nature. When deforestation was criticized, it was on the grounds that it was bad for mankind. But at the very least *it was beginning to be seen as a serious problem by some.* Even though the views and needs of these people differed from our own, we can locate in this period many of the disparate "sources" that have contributed to the making of sustainability, which in many ways is a conscious attempt to "return" not to pre-industrial society per se, but to a time when humans tread more lightly upon the Earth.

In Europe and its settler societies, the intellectual and cultural movements today known as the Scientific Revolution (late sixteenth to late seventeenth centuries) and the Enlightenment (late seventeenth to late eighteenth centuries) ushered in a period of intense interest in the natural world. There was no single event that triggered this widespread curiosity in nature and "natural philosophy," as science was called at the time, but rather a gradual development of disciplines gave shape and expression to that curiosity: chemistry, physics, mechanics, hydraulics, natural history (botany and biology), forestry, zoology, geology, anatomy, and so on. The new ideas associated with these disciplines eventually displaced the long-suffering Aristotelian paradigm of the natural world, which was buttressed by Galen's medicine and Ptolemy's astronomy and which had been endorsed and modified by the medieval Catholic Church. The followers of Aristotle throughout the Middle Ages had viewed the Earth as the center of the universe, assumed the existence of different kinds of matter, projected anthropomorphic meaning onto living organisms, and argued that the principles that governed bodies in outer space differed from those that governed the earthly realm.[13] But during the Scientific Revolution, a new paradigm came into existence that viewed the cosmos in mechanistic terms as something made up of motion and a single kind of matter.[14] There was only one set of physical laws that applied to both the earthly and heavenly spheres. And there was now a heliocentric universe that displaced poor old Earth from the center of it all.

This newfound interest in natural philosophy is evident in a wide range of practices and events in the seventeenth and eighteenth centuries, from the countless printed works of the period, including Isaac Newton's paradigm-shattering *Philosophiæ Naturalis Principia Mathematica* (1687) that helped establish the new physics, to the new academies of science that appeared throughout Europe and even in the New World, the most important of which were the Royal Society in London (1660) and the Paris Académie des sciences (1666).[15] It is also evident in the new journals dedicated to scientific exchange that appeared as well as the menageries, natural history cabinets (for anomalous and exotic specimens), and public displays of scientific devices that delighted crowds in eighteenth-century Europe.

An important aspect of this growing interest in science was that it brought with it a reevaluation of human beings and their relationship to the natural world. In *Man and the Natural World: Changing Attitudes in England, 1500–1800*, Keith Thomas discusses the deep-rooted Christian belief that the natural world had been created for the benefit of humankind. Nature, in a sense, belonged to human beings but needed to be pacified like a menacing enemy. However, in the seventeenth and eighteenth centuries, "some long-established dogmas about man's place in nature were discarded [and] new sensibilities arose towards animals, plants and landscape."[16] The new worldview that developed largely did away with the idea that humans were created and instead considered humans part of the "economy of nature." This is not to imply that anthropocentrism suddenly disappeared or that secular views suddenly eradicated religious ones. Even though humanity was now thought of, by many, as a part of nature, humans still held an exalted and dignified place within the natural order.[17] Nor did it mean a sudden admiration for nature on the part of most natural philosophers. René Descartes wrote of the need for humans to become "masters and possessors of nature."[18] Francis Bacon, the great propagandist of scientific utility in the seventeenth century, rallied humans to "conquer and subdue" nature, echoing the dictum in Genesis (1:28) to "replenish the earth and subdue it."[19] Likewise, the academies that Bacon helped dream up saw their role as creating knowledge that enabled the state to dominate nature. Indeed, Bacon spoke often of the relationship between "power" and "science."

Another consequence of this changing worldview was that it emptied the natural world of its magical or supernatural qualities and engendered the common belief that nature was inert and soulless. It was now an object of rational analysis rather than a source of awe and spiritual reckoning. Nature was something to be poked at, prodded, dissected, tortured for its secrets, and put on display. The dominant view was that there was no harm in destroying the natural world since it "felt" nothing and had no inherent rights or recognitions. Adam Smith, along with many others at the time, saw nature as "no more than a storehouse of raw materials for man's ingenuity."[20] For the natural philosophers of the eighteenth century—the remaining alchemists notwithstanding—the natural world was not an obscure and magical thing but rather something mundane and decipherable. For instance, the Baron d'Holbach, a radical atheist based in Paris, characterized nature as a giant unfeeling machine that could be understood through scientific analysis.[21]

The desire to study and make sense of nature is most apparent in the many taxonomies and natural histories that were produced in this period. The multivolume *Histoire naturelle* (1749–1789) of the great French naturalist Buffon, a catalog of all known animals and minerals, became the standard text for the biologists and geologists of that era.[22] The Swedish botanist and zoologist Carl Linnaeus invented modern binomial nomenclature and taxonomy in a series of publications between the 1730s and 1770s, and his system remains the structural foundation of the life sciences in our own day.[23] The first modern encyclopedia also came into being in this period. Denis Diderot and Jean le Rond d'Alembert edited a famous *Encyclopédie* (1751–1772; 17 volumes of text and 11 of plates), which solicited articles from dozens of contributing specialists and aimed to be a cutting-edge repository of all known knowledge.[24] By the end of the century, it seemed plain to many Europeans that nature had been fully exposed, categorized, and basically figured out.

From a certain perspective, then, it's not hard to see how the Enlightenment might have spurred on heartless environmental destruction and the expansion of a slave-based colonial empire, and that is certainly a reasonable argument to make. But the shift in worldview in this period or the pluralization of discourses also created the possibility for dissenting viewpoints. Many intellectuals and social observers used natural philosophy to *criticize* waste, degradation, social injustice, and illogical governmental policies. According to Richard Grove, "The growing interest in mechanistic analysis and comparison actually enabled rational and measured conservationist response."[25] Again, the divided legacy of the Enlightenment is made apparent.

One place to follow this clash of values is in the pages of Donald Worster's *Nature's Economy: A History of Ecological Ideas*. Worster employs two categories to describe competing views of the natural world in the eighteenth century: "Imperialism" and "Arcadianism." Linnaeus was the archetypal Imperialist who supported mankind's apparent domination over nature and viewed flora and fauna as little more than objects of dispassionate analysis. This was the prevailing view of the time. The Arcadians, by contrast, were exemplified by an English parson-naturalist named Gilbert White, who sought a "simple, humble life for man with the aim of restoring him to a peaceful coexistence with other organisms."[26] White saw harmony and complex systems within the natural world and clearly had a deep reverence for all living beings.

What's striking, though, is that both the Imperialists and the Arcadians contributed to the formation of ecological concepts— Linnaeus wrote of an "oeconomy of nature" and Gilbert White argued that "nature is a great economist," and thus both factions used the circulation of resources within human society as a metaphor for the cycles and systems within the natural world.[27] Moreover, both the Imperialists and the Arcadians at times mobilized knowledge of the natural world to counteract unsustainable environmental practices. Worster is therefore correct in arguing that multiple ecological viewpoints developed in the eighteenth century, well before "oecologie" became a term in 1866. The great outpouring of ideas and perspectives unleashed by the Enlightenment included many that valued the notion of humans living within their natural limits, even if many of the thinkers in this period cared little about nature in and of itself. This early ecological consciousness nonetheless became a powerful vantage point from which to attack human destructiveness and animosity toward the natural world.

In looking at the emergence of sustainability in early modernity, it becomes clear that the concept has roots in forestry. This is not a coincidence. In the period before the widespread use of fossil fuels, many world societies relied heavily on trees for fuel and other needs, and deforestation brought with it the specter of societal collapse. The forest was life sustaining, and because of the immediate relationship that pre-industrialized peoples had to this natural system, it was relatively easy to recognize its value and the effects of misuse. It is no exaggeration to say that, in Europe, the economy and social well-being were wholly dependent on the continued existence of woodland and a steady supply of forest resources. Urbanites and peasants needed wood to heat homes, build fences, and construct buildings. Without wood, most people would have frozen during cold winters, and the cooking of meals in most places would have been nearly impossible. Farmers and ranchers turned to woodlands for a wide range of needs, from nuts and berries that supplemented diets, to twigs and undergrowth that grazing animals used as fodder. Countless industries needed a steady supply of wood, too. Glue makers transformed tree sap into adhesives. Tanners and glassmakers and charcoal makers consumed vast quantities of timber and often contributed very directly to local deforestation.[28] European monarchies also valued woodland but for different reasons. For the elite, forests were important as hunting sites and places of leisure, but even more important was the fact that navies needed a constant supply of dense wood, especially oak and elm, to build armed ships. It took 2,000 to 3,000 suitable oaks to build a large warship.

Not only were forests recognized as vitally important, but there was a growing realization from the late seventeenth century onward that woodland was shrinking quickly, and that this was a problem with widespread consequences. The historical and ecological data show quite clearly that forests were, in fact, disappearing in this period. Per capita consumption rates stayed level in most places, but the European population grew considerably in the 1700s, creating new stresses on woodland resources. (In France, for instance, the population grew from 20 million in 1700 to 28 million in the 1790s.) Some wood-reliant industries expanded in this period, too. Coal began to be used more frequently in some parts of Europe in the eighteenth century, but its use seems to have slowed the rate of deforestation only late in the century or after 1800. In *Deforesting the Earth*, Michael Williams estimates that Europeans removed an astounding 25 million hectares of woodland and 40 million hectares of grassland between 1700 and 1850.[29]

We know that people from all social classes saw deforestation as a problem, but it was the literate elite who left the most detailed evidence of this incipient ecological consciousness. One of the first consciousness-raisers was John Evelyn, an English aristocrat and a founding member of the Royal Society. His 1664 *Sylva, or a Discourse of Forest-Trees, and the Propagation of Timber in His Majesty's Dominions*, deplored the woodland loss that plagued England in the seventeenth century. This best-selling book offered an encyclopedic description of tree species and contemporary practices for re-populating forests and planting, transplanting, pruning, and felling trees. It's also an impassioned plea for the nobility to replant and afforest the landscape.[30] Although Evelyn's main interest was maintaining naval and state power through energy independence, his book had a far-reaching impact on silviculture practices within and beyond the British Isles. His book inspired an act meant to refor-est the Royal Forest of Dean, for instance, and he motivated aristo-cratic landowners to plant "millions" of trees on estates throughout England.[31] Grober notes in Evelyn's book, "We are very close ... to the vocabulary of our modern sustainability discourse."[32]

A second "source" within forestry came from the administrative work of Jean-Baptiste Colbert, a French commoner who managed to climb the bureaucratic ladder to the post of Minister of Finances during the reign of Louis XIV. Colbert wanted to bring to forestry the same protectionist, corporatist management philosophy that he had brought to trade policies and guild regulation. In 1669, Col-bert organized and reformulated all the forest-related decrees that had been promulgated in France since the thirteenth century, is-suing an all-encompassing *Ordonnance sur le fait des Eaux et Forêts* that would govern forest policy until the nineteenth century.[33] This hugely influential forest code became the model of "rationalized" state forestry for governments all over Europe and beyond. How-ever, even though Colbert's decree might have slowed deforesta-tion in certain areas, its ultimate aim was to ensure enough wood for the navy and to protect the rights of forest owners (including the monarchy and the Catholic Church) against illegal poaching, grazing, and logging. As a result, the *Ordonnance* created more problems than it solved, triggering more violence and conflict be-tween peasants and state agents and facilitating inconsistent forest management policies, which enabled logging practices that drove deforestation.[34]

Given all this, Worster would probably categorize Colbert as an Imperialist—he was also an *actual* imperialist via his role in the mercantile-and-imperial French East India Company—since he cared relatively little about nature itself. Nature was merely a lump of inert matter meant to benefit humankind, and especially the needs of the state. Old-growth forests were protected because the crown needed them. What's more, the French monarchy actually accelerated the rate of deforestation by selling off royal forests, strengthening the power of the nobility to exploit forest resources, and encouraging forest clearance (for agriculture) in the 1760s, when the physiocrats entered the government.[35] As Grober notes, "On the eve of the Revolution in 1789, there was less woodland in France than in 1669."[36] Thus, Colbert's attempt at protecting forests and implementing a comprehensive forest management system met with mixed results in the early modern period, and the real importance in this code, for those interested in the history of sustainability, lies with the future treatises and codes that Colbert helped to inspire.

One of those forest treatises that owed quite a lot to Colbert was Hans Carl von Carlowitz's 1713 *Sylvicultura oeconomica, oder haußwirthliche Nachricht und Naturmäßige Anweisung zur wilden Baum-Zucht.*[37] In recent years, Carlowitz's innovative forestry manual has been recognized, quite rightly, as a watershed moment in the history of sustainability.[38] Carlowitz worked as a royal mining administrator in the Electorate of Saxony, one of the Germanic states that made up the Holy Roman Empire. In his early career, he traveled to France and England and became deeply influenced both by Colbert's "rational" approach to forest management and Evelyn's aristocratic appreciation for energy independence. As someone whose main concern was the vast and profitable mining industry in Saxony, where metallurgy and the mining of silver, copper, tin, and cobalt dominated the local economy, Carlowitz turned to the study of forests primarily because he understood that local industry relied on a large and constant supply of timber. He thus recognized the clear interrelationship between the economy and local natural resources and the fact that an industry could collapse if social authorities did not address and reverse the trend of deforestation.[39]

His aristocratic family had owned forests for generations, but Carlowitz began to educate himself in the science of forestry only late in his career. His famous treatise, *Sylvicultura oeconomica*, which was published a year before he died, represented several years of research and experimentation in Saxon forests. The wide-ranging book deals with such subjects as wood shortages in Saxony, inventions in metallurgy that could decrease fuel consumption, Roman forestry laws that prevented the cutting of young stands, natural causes of deforestation (storms, disease), human causes of deforestation (industry-driven overconsumption), the collecting of seeds, the possibility of regenerating forests, the importance of coppicing practices, and different methods of intensive forest cultivation. The upshot is that Carlowitz laid out the first comprehensive strategy for sustained yield forestry, and criticized the shortsightedness, mismanagement, and greed that had dwindled natural resources. In the words of Grober, "Carlowitz not only invents the word; he sketches out the entire structure of the modern sustainability discourse."[40] Here is one of the passages where Carlowitz states the importance of the "sustainable use" (nachhaltende Nutzung) of woodland:

> Thus the largest art, science, industry and institution based on it [forestry] exists in this region [Saxony]. As such, efforts must be made for the conservation and cultivation of wood. We must aim for a continuous, resilient, and sustainable use, because [forests] are an indispensable thing, without which the country and its forges could not exist.[41]

This seemingly self-evident statement was in fact a hard-earned realization that came about well after Germanic lands had entered a serious fuel crisis and as wood prices continued to climb. We shouldn't assume that woodland loss "caused" Carlowitz to fight back against deforestation, since many deforested societies around 1700 showed no interest at all in resource depletion, but it's nonetheless clear that the idea of sustainability occurred as a *reaction* to the mishandling of forests. Carlowitz's reaction, however, slowly became the dominant paradigm in forestry in Germanic countries, Western Europe, and North America.

Some of the ideas that Carlowitz defended were put into practice in the early decades of the eighteenth century: selective logging, coppicing, reforestation efforts, and the use of peat (and later coal) as an alternative to wood resources. In fact, *Sylvicultura oeconomica* became the standard guidebook for forest administrators. By the late eighteenth century, "sustainable yield forestry [had] developed into a science."[42] Deforestation rates slowed in the Holy Roman Empire as Carlowitz's ideas were expanded upon and implemented in universities and forestry schools in Harz, Zillbach, and Tharandt. In the early nineteenth century, universities in Saxony, Prussia, Nancy (Lorraine), and elsewhere were teaching sustained yield forestry, and Germans became the world's leading experts in this field.[43] Even in the late nineteenth and early twentieth centuries, students came from all over the world to study cutting-edge forestry techniques in Germany. Gifford Pinchot, the first Chief of the US Forest Service and a crucial figure in conservationism at the turn of the twentieth century, conducted his early training in Germany and other parts of Europe, bringing back to America the idea of sustained yield forestry.

Carlowitz, too, would probably fit into the category of Imperialist. Even though he makes the occasional comment about the beauty and "life-giving" qualities of nature, he viewed trees essentially as utilitarian resources.[44] The same could be said of Evelyn and Colbert, who saw nature as a collection of raw materials. But there is a second way to classify the ecological consciousness in the eighteenth century: between those who cared primarily about the needs of the state and its economic and political interests (Evelyn, Colbert, Carlowitz) and those who understood the complex value of trees and the social ramifications of woodland loss.

For instance, numerous French writers who participated in public essay competitions on trees and woodland loss—public writing competitions being an important intellectual venue in the Enlightenment—were able to identify the causes and effects of deforestation as well as some potential solutions to the problem. Although all of these essayists were fundamentally anthropocentric, they possessed a dynamic consciousness of how environmental destruction was linked to social and economic issues. Indeed, these early systems thinkers have left us with perhaps the strongest evidence that there was an incipient awareness of the three Es in the eighteenth century. Grober, who has written extensively about Carlowitz, argues that the Saxon administrator took interest in the "three pillars" of modern sustainability, because of Carlowitz's occasional reference to the poor.[45] However, I tend to see a greater focus on social justice in the writings of these French essayists, whereas Carlowitz cared about trees (environment) principally because they were an industrial instrument (economics) that ensured the continued wealth and power of Saxon elites (society).

The French essayists wrote first and foremost about the causes of deforestation. They connected the dots between woodland loss and the overconsumption of the industrial sector (especially metallurgy and charcoal-making), the greedy timber hoarding of the wealthiest classes in society, and illogical governmental policies that promoted land clearance to boost the agricultural sector. The final critique was aimed at the physiocrats, a group of free-market economists with close ties to Adam Smith, who exerted great influence on economic policy in France in the 1760s and 1770s, and believed in economic growth at the expense of natural resources.

As for the effects of deforestation, the essayists demonstrated a nuanced understanding of the social consequences of dwindling forest resources. They argued that deforestation caused the price of timber to rise—modern historians tell us that wood was 91% more expensive in 1785 than it had been in 1730[46]—and also noted the acute shortage of firewood experienced in many parts of France. Crucial here is the fact that they were highly cognizant that scarcity and high prices caused suffering, above all, for the poorest members of society. It was a social injustice to deprive the urban working classes of firewood that kept stoves and fireplaces ablaze. Even more surprisingly, these essayists went further than any contemporary bureaucrat in identifying what we would call today the "environmental" problems associated with deforestation. Numerous writers observed the deleterious effects that deforestation had on soil erosion. They noted that, when hillsides became denuded and root systems were removed, soil from higher elevations flowed downhill and ruined valley croplands. Countless farmers in provincial France had to cope with this issue in the late eighteenth century. They also discussed issues related to the water cycle, soil nutrients, and soil depletion.

Lastly, these essayists offered creative solutions to the problem of deforestation. They suggested the "conservation" of woodland—the term was used in the 1780s, well before the conservation movement of the late nineteenth century—wiser laws and administration of publicly owned forests, and sustainable forestry techniques, such as selective logging practices and coppicing. They also weighed the relative merits of privatizing forests as opposed to retaining a portion of French woodland as "commons" that would be shared by local inhabitants (a practice with deep roots in European history).[47] In short, these minor Enlightenment thinkers attacked environmental destruction on social and economic grounds. They were aware that the limits of acceptable consumption had been surpassed, and they reacted accordingly. To my knowledge, this is the best example that we have from the early modern period of Europeans criticizing unsustainable practices and conceptualizing society, the economy, and the environment as an interconnected system.

European societies weren't the only ones considering sustainable forestry measures at this time. In fact, many indigenous societies had used forests sustainably for millennia. One of the success stories comes from the highlands of New Guinea, where the local inhabitants have lived sustainably for 46,000 years, due in large part to an ingenious silvicultural practice of systematically transplanting and cultivating Casuarina seedlings that sprout naturally in riparian zones, thus ensuring a consistent supply of fuelwood.[48] In the Americas, aboriginal tribes from the Eastern seaboard of North America to the Amazonian rainforest used forests sustainably to meet their diverse needs. Small-scale slash-and-burn agriculture, strategic burnings in forests to trap game, and widespread agroforestry had been practiced successfully for long stretches of time before the arrival of Europeans and European ecological practices.[49] Indigenous agroforestry, in particular, was a sustainable land-use system in which groupings of trees and shrubs were planted close together in managed forests, which yielded fruits, burnable biomass, mulch, and other useful items.

In pre-industrial indigenous societies, these ecological practices tended to be bottom–up. In Japan, as with in Europe, the shift toward sustainable forestry was much more top–down, suggesting that global societies have taken different approaches to the sustainable use of land. During the Tokugawa Shogunate (1603–1867), Japan was unified under a single chief who oversaw a period of peace, prosperity, and growth. The population boomed, industry expanded, and cities grew. All of these trends put pressure on Japanese forests, which began to dwindle noticeably in the seventeenth century. Complicating the loss of natural resources was the fact that the Tokugawa opted to transform Japan into an isolated, self-sufficient society after some bad experiences with Portuguese Catholics who came to the island in search of trade and religious converts. After that, the only outsiders that the Japanese traded with (before the 1850s) were the Koreans and the Dutch, who were quarantined in special enclaves and harbors. In other words, no timber imports entered the country in this period. The Tokugawa began reacting to deforestation in the 1660s by urging people to plant seedlings, and by 1700 Japan had "launched a nationwide effort at all levels of society to regulate use of its forests" and to establish an "elaborate system of woodland management."[50] The Tokugawa regulated the use of wooded areas, gathered precise data on forest inventories (which ensured that the state acquired the best trees), implemented coppicing techniques, and began reserving some forest plantations for long-term sustainable logging.[51]

Little is known about the transmission of forestry knowledge between Europe and Japan in this period. Jared Diamond argues that Japanese laws developed "independently of Germany." But is it possible that Dutch mariners were the conduit for knowledge between the two cultural spheres? Did the Japanese influence the Europeans? Did the Europeans influence the Japanese? For the time being, it appears that the simultaneous development of sustainable silviculture in Japan and Saxony was merely a historical coincidence, and neither of these "advanced" societies seems to have taken much notice of the forestry practices of non-industrialized indigenous societies. It's debatable whether the Tokugawa might be considered Imperialists in Worster's sense of the word, and Diamond downplays the role that a "Buddhist respect for life" might have played in Tokugawa resource management strategies.[52] But what seems clear is that the Tokugawa enacted forestry laws in a top–down manner that centered on the needs of the state. In this

way, the Tokugawa seem closer in outlook to Colbert and Carlowitz than they do to the French essayists discussed above, since
their interest in trees had more to do with industrial economics
and politics than it did with concerns for social justice or the needs
of the poor.

It is also clear that the Tokugawa were very successful in their
approach to woodland conservation and consumption, and once
the outside world learned about their management strategies in
the late nineteenth century, Japan's became another globally recognized model of sustained yield forestry. It should certainly be
considered a "source" of the modern sustainability movement. On
balance, though, in the same way that Europe should not be seen as
an ecotopia, we must be careful not to overstate the sustainable consumption of Japan. Today, forests cover 80% of the country, which
is the highest percentage of forested area in any developed country,
even though Japan has a very high population density and a strong
appetite for timber. However, since the late 1950s, Japan has merely
exported its environmental degradation and deforestation by consuming great quantities of wood from tropical forests in Asia and
elsewhere.[53] Since about 1960, Japan has witnessed a sharp decline
in self-sufficiency in wood products, signaling a resource crisis that
has grown along with the Japanese economy.[54]

Governments in China, Taiwan, and western India in this
period also had top–down "forms of state forest conservancy, water
control and soil conservation . . . that were at least as sophisticated
as those emerging in the European sphere."[55] All of these contexts
are part of the diverse and polyglot history of sustainability. Evidence remains from China and Taiwan that scholars in this period
took a great interest in water and woodland conservation issues. In
India, the Amirs of Sind began afforesting over a million acres of
the Indus flood plain in the 1690s and created nearly 90 hunting
and forest reserves. When the British East India Company began

to rule the Indian subcontinent in the eighteenth century, it essentially adopted the practices of the local rulers.[56] Indeed, it is enlightening to see how often European colonial powers borrowed from non-Western cultures in imposing forest conservation measures that were often but not always necessitated by European colonial destructiveness. It is also instructive to see how widespread an interest in sustainable forestry had become in the eighteenth-century world.

A final example of sustainable forestry practices in the early modern period comes from the context of European colonialism, where a series of colonial bureaucrats influenced both by Enlightenment science and non-Western environmental practices began to reverse the trend of deforestation begun by first-wave European invaders. Richard H. Grove's *Green Imperialism: Colonial Expansion, Tropical Island Edens and the Origins of Environmentalism, 1600–1860* offers an excellent analysis of the complexities and paradoxes of what he calls "green imperialism." He argues that European-controlled islands in the Caribbean, the Atlantic Ocean, and the Indian Ocean had a major impact on the development of modern environmental consciousness and later the idea of "sustainable development." It was easier, in a sense, for European invaders to recognize environmental degradation and its consequences on isolated, self-contained tropical islands than it was back in the homeland. France, the Netherlands, and England had been slowly deforested over thousands of years, and this degradation seemed natural and tolerable by the early modern period. However, on islands, it was much easier to notice the rapid disappearance of forests, and its effect on climate, native species, and local hydrological systems, since these changes took place over a shorter amount of time and on a smaller morsel of land. Even though there had been awareness since antiquity of the effects of environmental degradation, it was not until the mid seventeenth century that a coherent and relatively organised awareness of the ecological impact of the demands of emergent capitalism and colonial rule started to develop, to grow into a fully fledged understanding of the limited nature of the earth's natural resources and to stimulate a concomitant awareness of a need for conservation.[57]

The irony here is that there would not have been a need for conservation had early European settlers not destroyed island forests— forests which had been used sustainably by local inhabitants for centuries or, in the case of uninhabited islands, had never dealt with an invasive species of homo sapiens. In the seventeenth century, colonial forces from England, France, the Netherlands, Portugal, and elsewhere took over such islands as Mauritius, St. Helena, Barbados, St. Vincent, Montserrat, and Jamaica in the hopes of transforming them into profitable agricultural outposts. Within a few decades, most of these islands had been badly deforested by a combination of land clearance policies, overconsumption of wood for industrial purposes, and the introduction of foreign livestock. Indigenous inhabitants, if there were any, were either killed off (and replaced by African slave laborers) or else severely diminished in population. In a process called "ecological imperialism," new flora and fauna as well as foreign agricultural practices overwhelmed and forever changed the ecology of these lands.

However, in the middle and late eighteenth century, a series of reformist bureaucrats wound up on these islands and sought to reverse the trends of environmental destruction. Many of them also opposed slavery, although as state-centered Imperialists, they cared first and foremost about the economic interests of their respective home countries, and their "environmentalism" (Grove's term) was rarely inspired by concerns for social well-being. Figures such as Pierre Poivre, who lived on French-run Mauritius, drew on Enlightenment science as well as Persian, Indian, and Zoroastrian botanical knowledge to protect old-growth forests. As Grove puts it, "Colonial states increasingly found conservationism to their taste and economic advantage, particularly in ensuring sustainable timber and water supplies and in using the structures of forest protection to control their unruly marginal subjects."[58] These technocratic bureaucrats realized that deforestation had economic and environmental impacts, too. It slowed agricultural output, since sugar production, for instance, required lots of wood; it caused streams to dry up or disappear, removing vital freshwater resources; and it caused devastating soil erosion.

Even more perceptive is the fact that these bureaucrats recognized that deforestation made local climates drier and warmer—rainfall seemed to diminish drastically in places that had been clear-cut—and it appears that these Europeans were some of the first people to realize that humans can, in fact, cause the climate to change.[59] Grove notes that "Poivre marshaled the climatic arguments against deforestation which had emerged by the 1760s [in Europe] and then persuaded the French colonial authorities of their importance."[60] Even though knowledge in this period of anthropogenic climate change differed from our own, this early concern for climate science is certainly part of the history of sustainability. The terrifying power of humans to change the climate has been recognized for over 250 years.

As a result of this growing environmental consciousness, European powers in numerous locations put in place strict policies meant to support the sustainable use of forests: St. Helena (late seventeenth century), the Dutch Cape Colony (late seventeenth century), Montserrat (1702), Mauritius (1760s–1770s), Tobago (1763), and so on. Thus, as we've seen in the case of Saxony and Japan, economic and political self-interest, combined with newly obtained forms of knowledge, were ultimately behind "green values." As odd and unsavory as it may seem, sustainability traces its roots primarily to imperialists (and Imperialists) who cared *very little* about nature or social justice and *very much* about state power, industrialization, and profit.

Finally, the emergence of sustainability is intrinsically tied to the creation of and reaction to classical capitalist economics in the eighteenth century. The term "economics" is actually a bit of an anachronism since the study of production, resources, wealth, and national income in this period was known as "political economy," a branch of moral philosophy. But what's key here is that the ideas of Adam Smith, Jean-Baptiste Say, Anne-Robert-Jacques Turgot, François Quesnay, the Marquis de Mirabeau, and other *économistes* had a major impact on changing economic models before 1800 and played midwife to the industrialization and growth that have been a part of the global economy for the past 250 years.

It is important to understand a bit about pre-1800 economic history for two reasons. First, many of the economists involved in recent ecological economics, such as Jeremy Rifkin, have harshly criticized Adam Smith and his economic theories. They believe that Smith's brand of capitalism should be "retired" from economics.[61] Second, it becomes clear that many of the concerns about classical economics that have been raised since the 1960s were first articulated by contemporaries of Smith and Turgot. In a sense, an acrimonious debate over the drawbacks and benefits of free trade, growth, and consumption has been raging since the eighteenth century. Beginning in the 1770s, critics began to appear who realized that a deregulated, growth-oriented economy brought potentially negative consequences for society, the economy, and the environment. These critics should be remembered as part of the history of sustainability.

So what is classical economics? The main idea is that a free market can regulate itself, bringing products and services to consumers and profits to producers and sellers without the active intervention of the state. In this sense, this idea was a reaction against the traditional, regulated economies that existed over all Europe, in which the state played a paramount role in dictating production and trade policies and supported artisanal guilds that controlled all aspects of a particular craft (prices, wages, output, quality, and so on). It is little wonder that Turgot abolished the guilds in 1776 while serving as the French minister of finances. In the same year, the Scottish moral philosopher Adam Smith published his profoundly influential *Wealth of Nations*, which quickly became the most emblematic declaration of this new economic philosophy. In it, Smith argues that the wealth of a nation is essentially the annual product of its labor. An increase in production was governed by an "invisible hand" via the free play of supply and demand that would, in theory, harmonize private self-interest (a vice) and public interest (a virtue).[62] He also lauds the importance of growth, the necessity of a strict division of labor, and the moral underpinnings of a free market (even though the state still played an important role in his system).

Smith, who was very much a child of the Enlightenment, was influenced in his economics by the theory of historical stages, which was popular in Scotland at the time, and which held that human civilization advanced along with new economic systems. Thus Smith associated capitalism with moral progress, which would later become a powerful idea during the nineteenth century. He was also greatly influenced by the physics of Isaac Newton, which he called "the greatest discovery that ever was made by man."[63] In the same way that Newton's Three Laws of Motion allowed for a self-regulating universe—minus the periodic intervention of God—a free market of goods and services based on supply and demand would allow for a self-regulating economy.

The other major contribution to classical economics came from the previously mentioned physiocrats in France, who had become an identifiable group of thinkers and bureaucrats by the 1760s. They fused natural law, medical theories, legal despotism, and agronomy to form a novel politico-economic theory based on market forces, the open circulation of goods within and beyond French borders, the sanctity of private property, limited taxation, an opposition to guilds, and the belief that all wealth ultimately derived from agricultural production.[64] "Physiocratie," which meant "rule of nature," became influential in France, and followers of this economic school, such as Turgot, Henri Bertin, and Clément-Charles François de L'Averdy, landed jobs as financial administrators within the French monarchy, enacting unprecedented liberal policies that induced nightmares for economic traditionalists. It was Vincent de Gournay, a noted physiocrat, who allegedly coined the phrase "laissez faire, laissez passer." Turgot, who became the most prominent member of this school, published his *Réflections sur la formation et la distribution des richesses* in 1766, which deals with agricultural production, free trade, and the division of labor, before serving as the top minister of finances in the 1770s. Turgot and Adam Smith met and became friends in 1766, exchanging ideas about how to transform the economies of Europe.

One aspect of classical economics that was controversial in the eighteenth century and which remains controversial today is the idea of economic growth. The scholar Anthony Brewer argues that growth was essentially a novel idea in this period and became one of the central tenets of this new school of economics. "Before the late eighteenth century, economic growth in its modern sense, that is, continuing growth in income and output over indefinitely long periods of time, was simply not on the intellectual map."[65] Another economic historian has shown that in the 600 years between 1086 and 1688, the English economy grew by only 0.29% per year on average but began to grow more steadily in the late eighteenth century as the Industrial Revolution intensified.[66] Smith and Turgot, in fact, rarely used the term "growth," preferring instead to speak of the "progress of opulence," the "continual increase of national wealth," and so on, but they agreed that capital accumulation was the chief engine of growth and that population would likely increase along with economic expansion.[67]

Traditionalists, moralists, and mercantilists attacked the idea of growth and deregulated trade, but even those who were not wedded to the old ways found fault in this new economic philosophy. The thing that bothered many people in the late eighteenth century was what growth apparently necessitated and implied. First, it tended to mean the privatization of publicly owned land. The phyisocrats, for instance, advocated for the "enclosure" of many forests and pastures that had traditionally been managed communally by village peasants.[68] Second, pro-growth policies, when adopted by governments, tended to result in the increased consumption of natural resources. As noted above, relatively few people cared about nature in this period, but they did care about resource shortages, soaring prices, and environmental degradation that jeopardized the livelihood of farmers. Third, craftsmen took exception to the economic and cultural changes ushered in by factories and new machines, and the reactive smashing of new machines had already become a common response to those changes by the 1770s. In short, the eighteenth century set the stage for an enduring conflict over the social, economic, and environmental costs of economic growth that would play out throughout the nineteenth century and down to the present day.

It was in France, above all, where the criticism of growth and its corollaries was the most fervent. Perhaps this is unsurprising since France went arguably further than any other eighteenth-century European state in experimenting with something other than a traditional, regulated economy, in which growth had essentially been a non-issue.[69] By the 1760s, it was commonplace in France to take aim at the "esprit de luxe" (loose translation: thirst for luxury) that had taken root in France, which was often associated with the economic policies of the physiocrats.[70] One essayist in the 1780s, echoing others around this time, denounced the deforestation that had occurred as a result of the physiocrats' land-clearance policies. "As a result," he wrote, "I would like a law that forbids the clearing of [forested] land with the most rigorous of penalties." The same essayist denounced the "disappearance of woodland" in France and lamented the loss of the "superb forests" that once covered the kingdom: "It hurts my eyes to see the enormous abuses that have caused this unfortunate situation, and which sooner or later will deprive us of this crucial natural resource." He goes on to denounce the physiocrats and the French monarchy for driving deforestation, which created timber shortages and price hikes that harmed the underprivileged most of all:

> In the southern regions of France, where wood is the rarest and the most expensive, the peasants can heat their homes only with what they can scavenge from the remaining communally held forests. This horrible scandal, the shame of those who have brought it about. . . . I denounce the administration—I denounce all of France. Many communities that I won't go to the trouble of naming have been left with little more than rocky soils that cannot support agriculture, and which once supported useful and necessary trees.[71]

The physiocrats, as with Adam Smith, certainly had many supporters, and I don't mean to imply that everyone in the run-up to the French Revolution of 1789 denounced the new economics. But at the same time, economic liberalism divided public opinion from the 1760s onward, and it is not difficult to find other writers in the late eighteenth century who recognized the true costs of economic growth.[72] Historians have only recently discovered that animosity toward greed, consumerism, growth, and what we would later call the Industrial Revolution was fairly widespread in this period. Moreover, at least some keen observers realized that economic policy had social and "environmental" consequences.

Finally, I will mention the influential philosophy and criticism of Jean-Jacques Rousseau, who was born in Geneva and spent much of his life bouncing around France, Switzerland, and elsewhere. By the 1760s, Rousseau had become a literary superstar throughout the Atlantic world, and his influence on intellectual life is felt to the present day. Although Rousseau wrote on many subjects, the three that have the most relevance for the history of sustainability are his contention that technological innovation did not make humans any happier or more virtuous, his critique of social inequality and its link to the natural environment, and his deep admiration for wilderness, rural values, and simple living. Rousseau was not necessarily a critic of Adam Smith or the physiocrats. Instead, his criticism of social and economic realities was much broader in scope, and he became the symbol for the naysayers in this period who looked askance at so-called social progress. What Rousseau desired was a movement back to the land and an emphasis on moral rather than technological or economic advancement.

His critiques of technology and social inequality appeared in separate essays that were submitted to academic prize competitions in the 1750s and were later published. His "Discourse on the Arts and Sciences" acknowledged that science and artisanal innovation had advanced in the course of human history, but rather than improving humanity, these advancements had actually weakened the moral capacities of humankind. He also anticipated the critics of the physiocrats by arguing that the eighteenth century had become an age of luxury, greed, and self-indulgence, to which Rousseau contrasted the virtuous simplicity of primitive peoples who lived among the trees.[73] The second essay, his "Discourse on the Origin and Basis of Inequality among Men," built on some of these assertions but shifted the focus to the semi-fictional (by his own admission) "state of nature" in which humankind had lived before the advent of civilization. Rousseau argues that humans are naturally equal, in a moral sense, and that inequality is little more than a fiction created by society.[74] Crucially, Rousseau links social inequality to the unequal ownership and usage of nature, initiating a line of argument that was later picked up by democratic revolutionaries, Karl Marx, ecological economists, and others. He argues that the ability to control and transform the natural world and natural resources is the true origin of social differentiation and, ultimately, poverty and human misery.[75] Through his frank and insightful critiques of technological "progress," luxury, consumption (which he always associated with cities), and social inequality, Rousseau was essentially an early opponent of the Industrial Revolution. His writings set the stage for later critiques of industrialization and growth that took off in the nineteenth century and in a sense remain relevant to the present day.

The third and final aspect of Rousseau's Arcadian philosophy that serves as a source of the sustainability movement is his admiration of nature. In works such as *Emile*, his great treatise on education, Rousseau extols the virtues of living and learning in a natural setting and even thinks of the love of nature as a kind of primal religion. Rousseau himself spent considerable time out in nature, which was a very unusual pastime before he made it popular. He traveled the back roads, hiked in the hills, lived in country cabins, and studied botany along the banks of alpine lakes. His love of nature differed from that of mainstream Enlightenment philosophers, who saw nature as an abstraction or an inert object of study. Rousseau's appreciation was more holistic and experiential. He thought that the creation of cities and civilized living created a wedge between humans and nature. This wedge inhibited humankind's innate and primitive moral capacities. To be virtuous was to live in harmony with the natural world. Unlike John Locke, who thought that nature had value only if humans mixed their labor with it, Rousseau saw nature as intrinsically valuable. As many historians have pointed out, Rousseau was a kind of early Romantic who, like the actual Romantics of the nineteenth century, considered mountains and forests and oceans as objects of beauty that provided the appropriate setting for personal and spiritual growth. It's hard to overstate Rousseau's impact on changing European attitudes toward nature. Even though it would be anachronistic to call him an environmentalist or even a sustainist, his fondness for nature, simplicity, and primitivism served as a major inspiration to future conservationists and environmentalists, and his writings provided a way out of a strictly anthropocentric worldview.

It is by now clear that sustainability became a loose theory in the early modern period, even if it lacked a standardized vocabulary. It's also clear that the sustainability movement of the present day owes a lot to the early pioneers of ecological thinking who were able to draw connections between human society, the natural environment, and the economy. Although this wasn't the refined systems thinking that took shape in the twentieth century, the critiques voiced in the eighteenth century set the stage for a fuller understanding of how human actions toward the environment have a rebounding effect on social well-being and economic reality. For Carlowitz and his contemporaries, sustainability wasn't yet a blanket critique of a particular mode of existence so much as it was a technical recalibration of governmental policy by a social elite with the training and influence to make that determination. It was an early and constructive reaction to a set of problems caused by economic expansion, population growth, and resource overconsumption. These advocates for "sustainable living" did not generate an ecotopia, but they did have limited successes, the most notable of which was in the domain of sustained yield forestry. They also became important inspirations for the detractors of industrialism in the nineteenth century and beyond.

Chapter 2

The Industrial Revolution and Its Discontents

The stock narrative of the Industrial Revolution (*ca.* 1760–late 1800s) is one of moral and economic progress. Indeed, economic progress is cast *as moral progress*. The story tends to go something like this: inventors, economists, and statesmen in Western Europe dreamed up a new industrialized world. Fueled by the optimism and scientific know-how of the Enlightenment, a series of heroic men—James Watt, Adam Smith, William Huskisson, and so on— fought back against the stultifying effects of regulated economies, irrational laws and customs, and a traditional guild structure that quashed innovation. By the mid-nineteenth century, they had managed to implement a laissez-faire ("free") economy that ran on new machines and was centered around modern factories and an urban working class. It was a long and difficult process, but this revolution eventually brought Europeans to a new plateau of civilization. In the end, Europeans lived in a new world based on wage labor, easy mobility, and the consumption of sparkling products. Europe had rescued itself from the pre-industrial misery that had hampered humankind since the dawn of time. Cheap and abundant fossil fuel powered the trains and other steam engines that drove humankind into this brave new future. Later, around the time that Europeans decided that colonial slavery wasn't such a good idea, they exported this revolution to other parts of the world, so that everyone could participate in freedom and industrialized modernity. They did this, in part, by "opening up markets" in primitive agrarian societies. The net result has been increased human happiness, wealth, and productivity—the attainment of our true potential as a species!

Sadly, this saccharine story still sweetens our societal self-image. Indeed, it is deeply ingrained in the collective identity of the industrialized world. The narrative has gotten more complex but remains *à la base* a triumphalist story. Consider, for instance, the closing lines of Joel Mokyr's 2009 *The Enlightened Economy: An Economic History of Britain, 1700–1850*: "Material life in Britain and in the industrialized world that followed it is far better today than could have been imagined by the most wild-eyed optimistic eighteenth-century *philosophe*—and whereas this outcome may have been an unforeseen consequence, most economists, at least, would regard it as an undivided blessing."[1] The idea that the Industrial Revolution has made us not only more technologically advanced and materially furnished but also better for it is a powerful narrative and one that's hard to shake. It makes it difficult to dissent from the idea that new technologies, economic growth, and a consumer society are absolutely necessary. To criticize industrial modernity is somehow to criticize the moral advancement of humankind, since a central theme in this narrative is the idea that industrialization revolutionized our humanity, too. Those who criticize industrial society are often met with defensive snarkiness: "So you'd like us to go back to living in caves, would ya?" or "you can't stop progress!"

Advocates of sustainability are not opposed to industrialization per se and don't seek a return to the Stone Age. But what they do oppose is the dubious narrative of progress caricatured above. Along with Jean-Jacques Rousseau, they acknowledge the objective advancement of technology, but they don't necessarily think that it has made us more virtuous, and they don't assume that the key values of the Industrial Revolution are beyond reproach: social inequality for the sake of private wealth; economic growth at the expense of everything, including the integrity of the environment; and the naïve assumption that mechanized newness is always a positive thing. Above all, sustainists question whether the Industrial Revolution has jeopardized humankind's ability to live happily and sustainably upon the Earth. Have the fossil-fueled good times put future generations at risk of returning to the same misery that industrialists were in such a rush to leave behind?

Narratives are everything, since they are never created spontaneously from "the facts" but are rather stories imposed upon a range of phenomena that always include implicit ideas about what's right and what's wrong. That is, they are inevitably moralistic.[2] The proponents of the Industrial Revolution inherited from the philosophers of the Enlightenment the narrative of human (read: European) progress over time but placed technological advancement and economic liberalization at the center of their conception of progress. This narrative remains today an ingrained operating principle that propels us in a seemingly unstoppable way toward more growth and more technology, because the assumption is that these things are ultimately beneficial for humanity.[3]

But what if we rethink the narrative of progress? What if we believe that the inventions in and after the Industrial Revolution have made some things better and some things worse? What if we have a more critical and skeptical attitude toward the values we've inherited from the past? Moreover, what if we write social and environmental factors back in to the story of progress? Suddenly, things begin to seem less rosy. Indeed, in many ways, the ecological and demographic crisis of the present day has roots in the Industrial Revolution.

For instance, consider the growth of greenhouse gases (GHGs) in the atmosphere since 1750. Every respectable body that studies climate science, including NASA, the National Atmospheric and Oceanic Administration, and the US Environmental Protection Agency (EPA), has been able to correlate GHG concentrations with the pollutants that machines have been spewing into the atmosphere since the late eighteenth century. These scientific bodies also correlate GHGs with other human activities, such as the clearing of forests (which releases a lot of carbon dioxide and removes a crucial carbon sink from the planet) and the breeding of methane-farting cows. But fossil fuels are the main culprit (coal, gas, and oil) and account for much of the increase in the parts per million of carbon dioxide in the atmosphere. The main GHGs, to be sure, are carbon dioxide (CO_2), methane (CH_4), nitrous oxide (N_2O), and a few others, many of which can be charted over time by analyzing the chemistry of long-frozen ice cores. More recent GHG levels are identified from direct atmospheric measurements.

What we learn from these scientific analyses is that the Industrial Revolution ushered in a veritable Age of Pollution, which has resulted in filthy cities, toxic industrial sites (and human bodies), contaminated soils, polluted and acidified oceans, and a "blanket" of air pollution that traps heat in the Earth's atmosphere, which then destabilizes climate systems and ultimately heats the overall surface temperature of the planet. The EPA is quite blunt about it: "Increases in concentrations of these gases since 1750 are due to human activities in the industrial era."[4] There it is: *human activities*—the same ones responsible for the "progress" of humankind. It's worth noting, too, that the population of the world only began to take off during the Industrial Revolution. For millennia, the population of homo sapiens was well below the 1 billion mark, until that number was surpassed around 1800. The world now has 7 billion people and counting. That's a lot of people who require food,

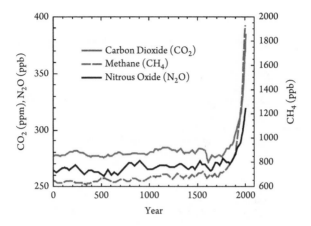

Figure 2.1. Carbon dioxide (PPM), methane (PPB), and nitrous oxide (PPM) in the atmosphere since 1750. Before the Industrial Revolution, CO_2 levels had long been stable at about 280 PPM. Now they're above 400 PPM. CO_2 levels have not been this high for at least 2 million years.

Source: USGCRP (2009). Cited online from the EPA's website: http://www.epa.gov/climatechange/science/causes.html.

energy, and housing and who place great strains upon global ecosystems. See Figures 2.1, 2.2, and 2.3.

When we take into consideration the damning evidence of these scientific analyses, the Industrial Revolution starts to look like something other than an "undivided blessing." It begins to look like, at best, a mixed blessing—one that resulted in technologies that have allowed many people to live longer, safer lives, but that has, simultaneously, destroyed global ecosystems, caused the extinction of many living species, facilitated rampant population growth, and wreaked havoc on climate systems, the effects of which will be an increase in droughts, floods, storms, and erratic weather patterns that threaten most global societies. The obvious question is, was it all worth it?

The picture gets even gloomier when we factor in global social inequalities. By many accounts, the Industrial Revolution actually

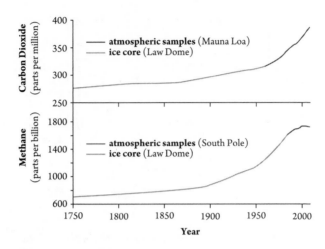

Figure 2.2. Carbon dioxide and methane levels in the atmosphere since 1750.

Source: NASA graphs by Robert Simmon, based on data from the NOAA Paleoclimatology and Earth System Research Laboratory.

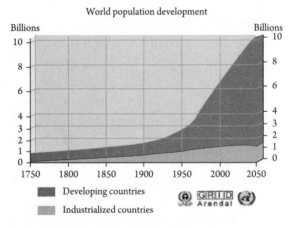

Figure 2.3. Population levels in developing and industrialized countries over time, with future projections.

Source: Philippe Rekacewicz, UNEP/GRID-Arendal.

made most people *poorer* and *more miserable,* while making a select few fabulously wealthy. As one historian has noted, steam-powered engines required fewer workers and thus resulted in higher unemployment—not to mention that the work was much dirtier and more dangerous than what it had been in the past. In 1812, the English county of Yorkshire had about 5,000 wool cutters (croppers) and apprentices. But by 1817 the number had dropped to 763. By the 1830s, the artisanal wool craft was basically dead in the area.[5] Even when the Industrial Revolution began to raise the overall wealth and standard of living in Britain, deep social inequalities remained in the country.[6]

Moreover, industrialization created deep and long-lasting social inequalities between Western Europe (and a few European settler societies) and the colonized world, which became poorer and more urbanized after 1800. The distinction between so-called developing countries and industrialized or developed countries began to take shape with the Industrial Revolution. Recent research conducted by UNICEF shows that global poverty rates are staggeringly high and that the income gap between "developing" countries and rich countries is wider than ever. "Under market exchange rates, we inhabit a planet in which the top quintile controls more than 80 percent of global income contrasted by a paltry percentage point for those at the bottom."[7] The economist Jeffrey D. Sachs adds that one-sixth of the world lives in extreme poverty and that poverty exacerbates environmental degradation and ethnic hostilities.[8] Poverty around the globe can't be blamed entirely on the steam engine, of course, but the Industrial Revolution nonetheless set in motion a series of economic and technological changes that have resulted in severe poverty and global inequalities.

All of this is to say that the simpleminded narrative of progress needs to be rethought. This is not a new idea: in fact, critics of industrialization lived throughout the Industrial Revolution, even if their message was often drowned out by the clanking sounds of primitive engines. In their own particular ways, thinkers and activists as diverse as Thomas Malthus, Friedrich Engels, the Luddites, John Stuart Mill, Henry David Thoreau, William Wordsworth, and John Muir criticized some or all aspects of the Industrial Revolution. The narrative of industrial-growth-as-progress that became *the* story of the period occurred despite their varied protestations. The Luddites questioned the necessity of machines that put so many people out of work. Engels questioned the horrendous living and working conditions experienced by the working classes and drew links between economic changes, social inequality, and environmental destruction. Thoreau questioned the need for modern luxuries. Mill questioned the logic of an economic system that spurred endless growth. Muir revalorized the natural world, which had been seen as little more than a hindrance to wealth creation and the spread of European settler societies around the globe.

Instead of providing a new metanarrative about the Industrial Revolution—one that is more negative than positive—this chapter will instead take a look at the discontents who became some of the first and most important critics of industrialized modernity and its effects. These figures have provided wisdom and intellectual inspiration to the sustainability movement. John Stuart Mill and John Muir, for instance, have each been "rediscovered" in recent decades, respectively, by ecological economists and environmentalists in search of a historical lineage. For the sustainists of the present day, it was these figures, and others like them, who were the true visionaries of the age.

Friedrich Engels may have coined the term "Industrial Revolution" in the 1840s, but it was the British economic historian Arnold

Toynbee who popularized it in the 1880s. Ever since then, there has been a raging debate among historians about whether the social, economic, and technological changes of the late eighteenth and nineteenth centuries constitute a proper "revolution." I tend to agree with Kirkpatrick Sale that the suddenness and momentousness of the changes warrants the use of term. The changes did not affect all parts of Europe, let alone the Atlantic world, at the same time or in the same ways, but at least in England (and much of Britain), this revolution

> changed not just the face of manufacturing but the places and purposes of production, the composition of the workforce, the character of the market, the patterns and sizes of settlements, and the role of families and communities; it happened rapidly, almost within a modest lifetime if we take the usual dates of 1785 to 1830, from the first successful steam-powered factory, inaugurating the age, to the first intercity railway, inaugurating another, during which the number of people in manufacturing exceeded that in agriculture for the first time in English history; and it was taken at the time for the cataclysm it was, a phenomenon of "wonder and astonishment [whose] rapidity . . . exceeds all credibility."[9]

What were those changes that cleaved history into two seemingly opposed eras, the industrial and the pre-industrial? There are different schools of thought on the Industrial Revolution, which focus, respectively, on technological changes, the birth of an urbanized working class, and economic growth—from slow, "Smithian" growth to rapid market growth—but none of these changes are mutually exclusive. Sale, for his part, has laid out the six most important transformations in England, which brought an end to pre-industrial society:

1. **The imposition of technology.** Coal-powered steam engines revolutionized the production of textiles and other goods and created the possibility of long-distance rail travel. Inventors engineered gas lighting and improved iron production. Machines came to replace human and animal labor, and much of the new technology ran on cheap and abundant coal, thus displacing the biomass, wind, and water that had powered the pre-industrial society. By 1850, Britain as a whole was extracting over 60 million tons of coal per year.[10]

2. **The destruction of the past.** Economic changes, combined with the enclosure and privatization of communal lands,[11] unraveled time-honored village traditions. The webs of duty and reciprocity that structured society were swept away by the new economic mode.

3. **The manufacture of needs.** The production of marketable goods created new "needs." Suddenly, the British had become consumers in a deregulated market. Cities grew precipitously in this period, as work shifted to urban centers, and the colonial world provided captive markets for the new surfeit of stuff.

4. **The ordeal of labor.** Many artisans became unemployed, and poverty rates soared to perhaps 30% in England and elsewhere. Work, for those who could find it, moved from the home or the shop to the factory.

5. **The service of the state.** The British government aided industrialization by creating a laissez-faire economy, which was in place by the late 1840s. They also supported industry though tax break and financial incentives and by maintaining order vis-à-vis the working classes.

6. **The conquest of nature.** "The Industrial Revolution was the first spectacular triumph of the human species over the

patterned, ancient limitations of the natural world."[12] The steam engine ran regardless of the weather or the season. Indeed, technology allowed Europeans to reach the long-desired goal of successfully dominating and "overcoming" the natural world.[13] Or so it seemed.

Sale's book, *Rebels Against the Future: The Luddites and Their War on the Industrial Revolution,* provides a convenient starting place for understanding some of the early resistance to these sudden transformations. The Luddites were a semi-organized "army" of artisanal wool workers who terrorized mechanized factories in Cheshire, Lancashire, Yorkshire, Derbyshire, and Nottinghamshire—the five counties at the heart of England—over a 15-month period in 1811 and 1812. They served under a mythical, entirely fabricated "leader" named Ned Ludd—a folksy stock hero—who lent his name to the movement. The Luddites were not opposed to all forms of modern technology, but they had little use for the water- and steam-powered wool-finishing machines that displaced the old hand-held ones, required fewer workers to operate them, and boosted production rates. In their own words, they opposed "Machinery hurtful to Commonality," since the new machine-based system for producing textiles in factories changed not only the mechanics of the wool trade but also dissolved traditional communal ties and drove artisans out of work or into treacherous factories.[14] Before the Luddites were mercilessly crushed by the army in 1813 through a series of skirmishes, mass arrests, show trials, and executions, these "victims of progress" managed to destroy over £100,000 worth of machines and garnered widespread sympathy for the moral challenge that they mounted against industrialization.

Ever since the 1810s, according to Sale, "Luddism has meant a strain of opposition to the domination of industrial technology and to its values of mechanization, consumption, exploitation, growth, competition, novelty, and progress."[15] However, the Luddites were not the first technophobes, nor were they the last. The practice of machine-breaking dated back to the late seventeenth century, and even after the demise of the Luddite movement, it remained a sporadic occurrence in Britain until about 1830, when working-class activism became more mainstream and institutionalized in the form of Chartism, which demanded political reform and universal male suffrage, and the trade union movement.[16] But because of their relative levels of organization, and the publicity that they received, the Luddites have gone down in history as the first serious opponents to modern technology.

What's so striking about the Luddites is that they directed their hostilities not toward capitalist factory owners or consumers but rather toward inanimate machines, which became hated symbols of the erosion of communal bonds and the advent of slave-like working conditions. As Lionel Munby noted some decades ago, "Technical progress itself was accepted [by the artisanal workers of the period], provided it was adapted for use by the hand workers themselves."[17] Even though the Luddites cared relatively little about the natural environment, their "critique" of the social and economic consequences of industrialization—expressed in the form of direct actions, pamphlets, and trial testimony—remains part of the background of the sustainability movement, even as the impulse to break machines has fallen by the wayside. As we will see in later chapters, sustainability is concerned in part with determining which technologies are beneficial for human society, the economy, and the natural world and which ones are not worth the costs. Sustainists are more highbrow than the Luddites, but at the end of the day, they're choosy about technology.

As mentioned, though, the Luddites were not the only critics of industrial change. The Romantic poets of the early nineteenth century waxed nostalgic about the loss of idyllic rural life and chronicled in verse the environmental changes that unfolded before their very eyes. All four of the major English Romantic poets— Byron, Shelley, Blake, and Wordsworth—denounced industrialism and sympathized with the plight of the Luddites. Wordsworth was particularly effective at recording the effects of industrial growth on the English countryside:

> The foot-path faintly marked, the horsetrack wild,
> And formidable length of plash lane. . . .
> Have vanished—swallowed up by stately roads. . . .
> And, wheresoe'er the traveller turns his steps,
> He sees the barren wilderness erased,
> Or disappearing.[18]

Indeed, Romanticism was a kind of literary rebellion against the Industrial Revolution. From Wordsworth to the German Goethe, the Romantics viewed the natural world as an organic whole whose integrity had been compromised by new industrial threats: coal mines, train lines, belching factories, polluted waterways, expanding cities, and the endless roads that trampled over the countryside. By contrast, the Romantics rejected the notion that nature was a lifeless storehouse of raw materials or a dumping ground and considered it instead a source of nourishment for the soul.[19] The "progress" of industrialism was, on the contrary, a regrettable sadomasochism that hindered spirituality and humankind's capacity to live harmoniously with the natural world.

The most famous back-to-the-land Romantic of the period was the American naturalist Henry David Thoreau. Growing up in Massachusetts in the 1820s and 1830s, Thoreau had witnessed the arrival of industrialism in North America and soon became a critic of the environmental wreckage and consumerism that resulted from it. After studying at Harvard, he opened a small school and

befriended Ralph Waldo Emerson, the American transcendentalist writer.[20] (Transcendentalism refers to a counter-cultural intellectual movement in 1820s and 1830s New England that adored
nature, despised the corruption in religion and politics, and advocated for peaceful self-sufficiency.) Between 1845 and 1847, Thoreau lived in a rural, microscopic cabin near Walden Pond in his
home state. His famous book of 1854, *Walden: Or, Life in the Woods*,
details his personal experiment in simple living.[21] While living in
the woods, Thoreau gained an appreciation for solitude, natural
harmony, and "self-reliance," which was also the name of a famous
essay by Emerson.[22] In "Winter Walk" and *The Maine Woods*, he
extolled the virtues of hiking and conveyed an "Arcadian" conception of ecology.[23] He also developed ideas about civil disobedi-
ence, for which he became rather well known in his own lifetime,
and rejected the modern luxuries and consumerism embraced by
many of his peers in New England.[24] His bucolic self-sufficiency,
his awareness of growth-driven deforestation, and his denial of
urban materialism made Thoreau the darling of the early conservationists and more recently of environmentalists, sustainists, and
anyone who values simple living.[25] As one observer has noted, "At
the foundation of modern sustainability lies the human connection
with nature, expressed first in America through the New England
transcendentalist movement."[26]

Back in Europe, some of the fiercest opposition to laissez-faire
industrialism in the 1830s and 1840s came from working-class
movements: trade unions, rights activists, and socialist parties. One
of the principal causes of the revolutions that swept across Europe
in 1848 was the demand on the part of the working class for higher
wages and better working conditions in factories. But even in countries that somehow avoided the upheavals of 1848, such as England,
workers organized against the new economy and its damaging consequences. What does the modern sustainability movement take

from the leftist critique of the Industrial Revolution? One could identify at least three themes:

1. **The critique of social inequality.** It tended to be those on the left who took issue with the social stratification worsened by the new economy. In the socialist tradition, which had "utopian" and "scientific" branches, figures such as Henri de Saint-Simon, Charles Fourier, Robert Owen, Pierre-Joseph Proudhon, Louis Auguste Blanqui, Karl Marx, and Friedrich Engels championed the notion of the natural equality of humankind, and poured vitriol on the propertied classes (the bourgeoisie or capitalists) who grew wealthy off the labor of the working poor (the proletariat).

2. **The critique of the centralization of resources and power.** Although twentieth-century communism is associated with economic and political centralization, early leftists railed against the concentrated wealth of the capitalist class—always a very small population of factory owners, resource managers, and financiers—and the excessive powers of government officials who backed industry and often profited from it. By contrast, they advocated for a wider and more just distribution of wealth, resources, and authority.

3. **The realization that free trade and economic growth led to environmental degradation**, which negatively affected the health, safety, and standard of living of the working class. Although few nineteenth-century leftists were explicitly "environmental" in orientation, some devised a holistic or ecological critique of industrial life, which centered, of course, on the tribulations of the proletariat.

The best example of this ecological critique of the Industrial Revolution is Friedrich Engels's 1845 *The Condition of the Working*

Class in England in 1844. Engels was born in Prussia to a well-to-do family involved in cotton manufacturing. As a young man, he bounced around Europe working as a journalist and getting involved in radical politics. Not long after meeting Karl Marx, who would become a lifelong friend and collaborator, the budding socialist was sent by his father to the family textile factory in Manchester, where exposure to some real-world capitalism was meant to set the young man straight. The plan backfired badly. The living and working conditions in Manchester only hardened Engels's radicalism, and he began to research and write about the plight of English workers. The book that he published in 1845 is a devastating condemnation of industrialism that explodes the myth of human progress. His sharply worded critique connects the economic changes of the period to the "social murder" of the working class—the way in which laws, norms, and structures eventually overwhelm and kill off workers—and the despoliation of the natural environment. He attacks the way in which property has been centralized "in the hands of the few" and the fact that "population becomes centralized just as capital does [in large industrial cities]."[27] Engels describes in vivid detail the filthy working-class towns surrounding Manchester: "The towns themselves are badly and irregularly built with foul courts, lanes, and back alleys, reeking of coal smoke, and especially dingy from the originally bright and red brick, turned black with time, which is here the universal building material."[28] Engels also drew on government reports to analyze the water and air pollution that came from industrial sources and which created a health crisis, above all, for the poor. Here is one of his famous passages on the industrial sludge in area waterways:

> At the bottom flows, or rather stagnates, the Irk, a narrow, coal-black, foul-smelling stream, full of debris and refuse, which it deposits on the shallower right bank. In dry weather, a long

string of the most disgusting, blackish-green, slime pools are left standing on this bank, from the depths of which bubbles of miasmatic gas constantly arise and give forth a stench unendurable even on the bridge forty or fifty feet above the surface of the stream.[29]

In the short term, Engels's book did not have the impact of *The Communist Manifesto* (1848), which he co-authored with Marx and which became an extremely influential call to arms for the workers of the Industrial Revolution. But, in the long run, Engels's work set a precedent for addressing the costs of an industrialized economy: pollution, poverty, ghettoization, child labor, and living conditions not worthy of human dignity. Of course, we are still dealing with the costs of an industrialized economy, from the off-the-charts air pollution in Beijing, to the oil-soaked communities in the Niger Delta, to the Bhopal industrial disaster in India (1984), and the rising atmospheric concentrations of GHGs. Even though sustainability assumes the existence of a capitalist framework, the movement takes from the far left an acute awareness about how industrial society can exacerbate social inequalities and poison the environment. Engels, in particular, is remembered today as one of the first writers to raise awareness about urban unsustainability and occupational health issues that affected industrial workers and their children.[30]

To be sure, though, sustainability is also informed by thinkers at the other end of the political spectrum: social conservatives and classical economists, all of whom were fundamentally supportive of industrial capitalism. No doubt many of these thinkers would be shocked to discover that today they're associated with something called "sustainability" and that they're mentioned in the same breath as Friedrich Engels and the Luddites. But part of what makes sustainability so fascinating is that it draws on a diverse set of political viewpoints. It thus has roots in the work of the decidedly *not* left-wing political economists Thomas Malthus, John Stuart Mill, David Ricardo, and William Stanley Jevons, who are frequently cited in the literature on sustainability as early critics of consequence-free industrial growth.[31] Collectively, they had deep-seated reservations about the ability (or even desirability) of growth—of cities, of natural resource consumption, of population—to continue indefinitely. Even though these economists helped establish the classical theory of capitalism, they dissented from the pie-in-the-sky optimism of many of their contemporaries. For this reason they have been appropriated by the sustainists of the past few decades.

Thomas Malthus, the British political economist, demographer, and reverend, drew on the fledgling field of population studies to question the zeal for limitless growth and societal progress touted by William Godwin, the Marquis de Condorcet, and other Enlightenment optimists. His cautionary, even curmudgeonly, book of 1798, *An Essay on the Principle of Population*, came across as misanthropic to a culture enamored by technological advancement. He argued in the most sobering of terms that economic growth tended to increase the population at a rate that was unsustainable. His contention was that increases in food production could rise, at best, at an arithmetical rate, whereas population can rise at a geometric rate, leading to an eventual demographic crisis. "It may safely be pronounced, therefore, that population, when unchecked, goes on doubling itself every twenty-five years, or increases in a geometrical ratio."[32] But Malthus also spoke of "checks" on the population, which prevented overpopulation and which he divided into two classes: "positive checks" and "preventive checks." Positive checks were involuntary, unwanted events, whether man-made or environmental in origin, that tended to have an adverse effect on human reproduction. Examples included famine, disease, wars, plagues, severe labor, inclement weather, poverty, crowded towns, ineffectual nursing, and "excesses of all kinds."[33] Preventive checks, by contrast, were rationally calculated decisions made by humans to avoid reproduction. They included fornication, the avoidance of marriage, "unnatural passions" (i.e., homosexuality), birth control, and other forms of family planning, most of which the prudish Malthus found morally objectionable.[34] His central message, though, was that something always prevents population from getting too large, and society is ethically obliged to face this challenge and do what's necessary and morally appropriate to avoid future catastrophe.

It is debatable whether Malthus was the first European political economist to recognize that economic and demographic growth cannot (or should not) continue indefinitely. For instance, the highly regarded historical demographer E. A. Wrigley has argued that Adam Smith, along with Malthus, Mill, and "all the classical economists," assumed that "economic development must tend toward a condition that came to be termed the stationary state. Smith's views are visible in the passage [in *The Wealth of Nations*] ... in which he refers to the concomitants of the achievement of the 'full complement of riches' that would be the ultimate fate of most economies."[35] But even if Malthus was not the first to suggest that economies could not grow indefinitely, his "ecological" approach to this argument has certainly been the most influential since the early nineteenth century.[36] Indeed, it is difficult to overstate the impact that his *Essay* has had on economists, systems theorists, demographers, and ecologists. The ecological economists of the 1960s and 1970s, for instance, took Malthus's warnings as the starting point for their critique of economic growth and its consequences. The importance of Malthus is glimpsed in the Club of Rome's 1972 *The Limits to Growth*, in Herman Daly's 1977 *Steady-State Economics*, and in E. J. Mishan's 1977 *The Economic Growth Debate: An Assessment*.[37] Collectively, these "Neo-Malthusians" argued for a new approach to economics that took ecological limits more seriously and allowed society to avoid the "doomsday" scenario laid out by

Malthus.[38] Even today, discussing the ways in which environmental and demographic issues destabilize society is still known as a "Malthusian debate."[39]

This debate has become even more polarized in recent years. Those who favor economic and demographic growth and those who tend to see the Industrial Revolution as an "undivided blessing" consider Malthus to have been needlessly cynical. Joel Mokyr, for instance, agrees that "population pressure prevented income per capita from growing" in the years *before* the Industrial Revolution, but, he contends, in the nineteenth century, "modern economic growth consisted of overcoming these demographic negative feedbacks. The significance of the Industrial Revolution was that the race between babies and resources was won, resoundingly, by resources."[40] Indeed, many historians have argued that Malthus was essentially wrong because he didn't understand the redeeming qualities of modern technology, which, in the long run, increased food yields, lessened the impact of many of the "checks" mentioned above, and improved the standard of living in many parts of the world.[41]

However, those who see the Industrial Revolution as ushering in an age of pollution, overconsumption, and overpopulation and those who denounce growth as ecologically destructive and unsustainable tend to see Malthus as enduringly relevant. For instance, Diamond's *Collapse* includes a chapter on the Rwandan Genocide of 1994 that is entitled "Malthus in Africa: Rwanda's Genocide."[42] In it, Diamond argues that Malthus's basic model makes sense in East Africa. Famine and overpopulation—Rwanda and Burundi are the two most densely populated countries in Africa—exacerbated ethnic tensions and led to the slaughter of nearly 1 million Rwandans.[43] Malthus's model thus clearly remains useful in certain contexts. What's more, the Neo-Malthusians and other advocates for sustainability argue that Malthus's basic contentions were valid; he was just wrong about the time frame. The main argument of *The*

Limits to Growth is that the world would likely bump up against Malthusian limits at some point in the mid-twenty-first century.[44] In the meantime, humans could either sit around and wait for the "positive checks" to take effect or enact a series of "preventive checks" to avoid a mass die-off of our species. That is, no matter how great a society's technology, ecological limits can *never* be overcome or negated but merely staved off.

It wasn't just Malthus in this period who raised concerns about the ecological limits to growth. The financier, member of parliament, and political economist David Ricardo published an influential work in 1817 called *On the Principles of Political Economy and Taxation*, in which he argued that the fixed quantity of arable land in a country—or on the entire globe—would eventually hamper economic and demographic growth. Ricardo crafted the theory of declining marginal returns (the law of diminishing returns), which holds that, as high-quality land becomes scarcer, rents will begin to climb, labor costs will rise, and food will become more expensive, thus limiting profitability and future investments in the economy.[45] As E. A. Wrigley noted, "Malthus, Smith, and Ricardo, therefore, shared the conviction that economic growth must be limited because the land (in a literal and narrow sense) was a necessary factor in almost all forms of material production, and the supply of land was virtually fixed."[46] But Ricardo's warnings were not meant to be a critique of capitalist economics, and in fact Ricardo, along with William Huskisson, was one of the principal champions of laissez-faire economics in early-nineteenth-century England. As a result, Ricardo is one of only a handful of political economists who is still invoked favorably both by neoclassical and ecological economists. Lester R. Brown, who would rank among the latter, draws on Ricardo in his work on "building a sustainable society."[47]

A second political economist who is still cited by neoclassical and ecological economists is William Stanley Jevons, who became well known later in the nineteenth century for his expertise on coal and for his mathematical approach to economic analysis. In his 1865 book, *The Coal Question; An Inquiry Concerning the Progress of the Nation, and the Probable Exhaustion of Our Coal-Mines*, Jevons did the unthinkable and addressed the prospects of a United Kingdom without ample supplies of coal. Even though coal was cheap and abundant in his day—Jevons estimated that England alone produced 80 million tons of coal in 1860—he recognized that the "growing rate" of coal consumption would hasten the disappearance of this "incredible" fuel. He also rejected the conventional wisdom of the day by arguing, counterintuitively, that improvements in the efficiency of fuel-consuming machines would lead to *increases* in fuel consumption since consumers would use savings to buy more fuel, and producers would have more resources lying around waiting to be sold: "As a rule, new modes of economy will lead to an increase of consumption according to a principle recognised in many parallel instances."[48] This insight has since been labeled the "Jevons Paradox" and remains hugely influential in the economics of sustainability.

For instance, Herman Daly has argued that efficiency is a false panacea for overconsumption since it ends up backfiring and driving resource depletion.[49] More recently, the sustainist Tim Jackson has picked up on the same idea to argue against the notion that one can "decouple" economic growth from "material and energy requirements"; new technologies can actually exacerbate resource exhaustion rates and contribute to societal overshoot.[50] Similarly, the resource economists Mathis Wackernagel and William Rees have dealt with the Jevons Paradox in one of their essays on natural capital and ecological footprints. In it, they argue that any cost savings from efficiency gains should be "taxed away or otherwise removed from further economic circulation. Preferably they should be captured for reinvestment in natural capital rehabilitation."[51]

Thus, Jevons remains a crucially important figure in cautioning against the naïve assumption that efficiency alone can prevent over-consumption of resources.[52]

The final classical economist of the Industrial Revolution to mention is probably the most important: John Stuart Mill. Mill grew up in an intellectually intense environment, tutored by his overbearing father, James Mill, who was a utilitarian, and surrounded by Jeremy Bentham—the founder of utilitarianism—and the economists Jean-Baptiste Say and the aforementioned David Ricardo. Mill went on to become a highly regarded political economist, theorist of liberalism, and advocate for women's rights. His defense of freedom and free speech in *On Liberty* remains standard reading for university undergraduates. Of equal importance is his economic masterpiece, the *Principles of Political Economy* (1848), which offered a comprehensive theory of capitalist economics and which developed and refined the ideas of Smith and Ricardo, both of whom Mill held in high esteem.[53]

Why do ecological economists and other sustainists turn to Mill for insight and inspiration? Simply put, because of his belief that a growth-bound economy would—and should—ultimately culminate in a mature and prosperous state. Instead of endless growth, Mill envisioned a day when the economy would level off in terms of wealth creation, throughput, consumption, and population. Mill discusses this idea in a chapter called "Of the Stationary State":

> When the progress ceases, in what condition are we to expect that it will leave mankind?
>
> It must always have been seen, more or less distinctly, by political economists, that the increase of wealth is not boundless: that at the end of what they term the progressive state lies the stationary state, that all progress in wealth is but postponement of this, and that each step in advance is an approach to it.[54]

He also shared Malthus's misgivings about the tendency for population to grow at an unsustainable rate: "Even in a progressive state of capital, in old countries, a conscientious or prudential restraint on population is indispensable, to prevent the increase of numbers from outstripping the increase of capital, and the condition of the classes who are at the bottom of society from being deteriorated."[55] In the process, Mill distances himself from those contemporary political economists who advocated for endless growth in the "progressive state":

> I cannot, therefore, regard the stationary state of capital and wealth with the unaffected aversion so generally manifested towards it by political economists of the old school. I am inclined to believe that it would be, on the whole, a very considerable improvement on our present condition. I confess I am not charmed with the ideal of life held out by those who think that the normal state of human beings is that of struggling to get on; that the trampling, crushing, elbowing, and treading on each other's heels, which form the existing type of social life, are the most desirable lot of human kind, or anything but the disagreeable symptoms of one of the phases of industrial progress.[56]

Finally and crucially, Mill argued that economic growth and development were manifestly *separate* from moral and cultural development. He thus rejected a central tenet of the industrial age that assumed that human society could only advance together with economic expansion. Mill separated these two principles and argued that a stationary state would allow for greater focus on improving human culture.

> It is scarcely necessary to remark that a stationary condition of capital and population implies no stationary state of human

improvement. There would be as much scope as ever for all kinds of mental culture, and moral and social progress; as much room for improving the Art of Living, and much more likelihood of its being improved, when minds ceased to be engrossed by the art of getting on.[57]

It is remarkable to learn that many of the classical economists and, above all, Mill rejected an economic system that would expand indefinitely. It is pretty clear that business-as-usual, pro-growth, neoclassical economists of the twenty-first century have overlooked this crucial aspect of capitalism as it was originally conceived. Even though Mill was not the most explicitly ecological thinker—he cared far more about social and economic issues—his defense of the stationary state plays a vital role in the economics of sustainability. Systems theorists and resource economists have recast Mill as an early defender of an economic system that respected ecological limits and valued human development over the quest for "more." As Mill himself put it, "It is only in the backward [impoverished] countries of the world that increased production is still an important object: in the most advanced, what is economically needed is a better distribution."[58]

However, as some historians have argued, Mill's conception of a stationary state, rooted in stability and the just distribution of wealth, essentially fell by the wayside during the great age of economic growth in the late nineteenth and twentieth centuries, only to be rediscovered in recent decades, as our financial systems falter and our ecological crisis grows.[59] Tim Jackson contends that Mill was the first economist to say that more money and more stuff does not necessarily make the individuals in a society any happier or more fulfilled.[60] Herman Daly's influential theory of the "steady-state economy" is directly inspired by Mill's stationary state; indeed, one could even call Daly a Neo-Millian. As we will see in chapter 4,

Daly used Mill in the 1970s to argue in favor of a nongrowth economy based on a regulated population, a stabilized use of resources, and a fair distribution of resources, income, and wealth.[61]

Clearly, the economics of sustainability traces its origins back to the Industrial Revolution. And, to be sure, criticism of industrialism came from all along the political spectrum. It might have been the left who had the strongest opposition to the new economy, but the classical economists laid much of the groundwork for theorizing the ecological limits to growth. The sustainability movement thus draws from social conservatives (such as Malthus), capitalist liberals (such as Mill), and socialists (such as Engels) but weaves diverse viewpoints into a coherent philosophy of wisdom and endurance. In fact, the economists of the mid-nineteenth century seem to have exerted more of an influence on modern ecological economics than any of the works that appeared between the late nineteenth century and the 1950s.[62] But the main point is that the critique of pollution, inequality, and unsustainable growth in the nineteenth century remains influential and informative for sustainists in our own times.

Equally important are the scientists and naturalists of the nineteenth century, to which the modern sustainability movement owes as much as it does to that century's political economists. It was in this period that the natural sciences developed into specialized fields and moved out of scientific academies and into colleges and universities. There is much to say about the development of physics, chemistry, and other sciences in the nineteenth century, but the remainder of this chapter will focus on the growth and maturation of the field that has undoubtedly had the greatest impact on the concept of sustainability: ecology. The obvious place to begin is with Charles Darwin's paradigm-shattering, epoch-making theory of evolution, which appeared in his 1859 *On the Origin of Species* and which has had a major influence on modern conceptions of ecology in addition to its more well-known impact on biology.[63] Indeed, it is no coincidence that the word "oecologie" was coined in 1866 by a devoted follower of Darwin named Ernst Haeckel, who was looking for a term to encompass the study of the natural conditions of existence.[64] Even though Darwin's writings on evolution had relatively little to do with the Industrial Revolution, the science of ecology that he helped establish has been the basis since the nineteenth century for understanding how humans can live within their means, and, as such, it has often been at loggerheads with industrialism.[65]

Charles was the son of Erasmus Darwin, the eminent naturalist and member of the Lunar Society of Birmingham, an Enlightenment club that brought together natural philosophers and industrialists. As a young man, Charles was exposed to Erasmus's poetic musings on evolution, which may have played some role in Charles's view of the cosmos. He was also exposed to an aggressively competitive industrial society, which might have influenced his view that natural relationships were defined largely by competition. For nearly five years between 1831 and 1836, the young Darwin sailed around the globe on the HMS *Beagle*, serving as the live-in naturalist on a naval ship involved in surveying the coastlines of South America and various islands. It was in the course of these voyages that Darwin developed his theory of evolution—the idea that species evolve through a process called "natural selection," in which some members of the species are more successful than others in the endless "struggle for existence" or what Herbert Spencer later called the "survival of the fittest."[66]

Darwin's theory essentially scrapped the older ecological model that had reigned for much of the nineteenth century and which was most closely associated with Alexander Von Humboldt, the German scientific adventurer whose Romantic belief in the "great harmonies of Nature" was eclipsed by Darwin's disharmonious idea of perpetual struggle and adaptation.[67] In crafting his theory, Darwin drew on Charles Lyell's *Principles of Geology*, which dealt with geological "struggles" and transformation, and Thomas Malthus's cautionary writings on population.[68] Darwin lifted from Malthus the idea that a species—human or otherwise—can flourish only if it avoids overharvesting its food sources. That is, both Darwin and Malthus were aware of the fact that organisms must live within their natural limits and that overstepping those bounds will result in the collapse, decline, or even extinction of a species.[69] This foundational idea remains hugely relevant to ecological economists and sustainists down to the present day, even if ecologists of the present day recognize a range of relationships in nature, from the competitive to the symbiotic.

Darwin, it goes without saying, had a massive influence on the life sciences and set the agenda for ecological debates for decades after his death. Even though his "tree of life" metaphor—the idea that all beings are organically connected—was borrowed from the Romantics' lexicon, his approach had the effect of distancing natural science from the Humboldtian conception of balance and stability.[70] Darwin's work also had the effect of weakening anthropocentrism and especially Christian creationism, which posited inter alia that the universe existed to meet human needs.[71] After Darwin, the natural world seemed menacing and mutable, and humanity was demoted, in a sense, to the level of any other organism struggling to survive. Although Darwin treated humans as a unique species in his 1871 *The Descent of Man*, he theorized that all species lived in a web of interconnected existence and that survival was based upon the ability to adapt to and thrive within the natural environment. Darwin's ecological ideas and theory of evolution directly influenced the conservationists of the late nineteenth century and the environmentalists of the twentieth century, and his theories remain part of the background of sustainability.

Following Darwin, George Perkins Marsh was one of the first thinkers of the nineteenth century to draw on ecology as a tool for understanding and criticizing human destruction of the natural world. Marsh might not be a household name, but he was an important environmental writer in the late nineteenth century, and his legacy has been redefined in recent years. Indeed, Marsh has even been called "American's first environmentalist."[72] Marsh was a New England patrician and man of letters who gained election to the US House of Representatives before serving as a diplomat in Italy and the Ottoman Empire. While living in Italy in the 1860s, he became interested in forestry and ecology and began studying the environmental changes that the Italian peninsula had experienced since Roman times. His research culminated in his much-vaunted work of 1864, *Man and Nature*. According to one source, *"Man and Nature* was not an instant best seller, but over time it became known as an important book and sold well."[73] What readers found in the book was a sharply worded ecological condemnation of humankind's "reckless destructiveness" of the natural world.[74] Marsh drew on the work of both Humboldt and Darwin and, in fact, preferred the former's ecological model of balance and equilibrium to the latter's model of struggle and adaptation.[75] Marsh's main critique of Darwin's ecology was that it minimized the role that humans played in shaping natural processes.[76] His contention that humankind had become a destructive force of nature helps explain why he revised and republished his book in 1874 under a new title: *The Earth as Modified by Human Action*.

In the original edition, Marsh discusses such topics as the ecological wisdom of native peoples, the important functions played by forests (what we would now call "ecosystem services"), the ways humans affect the climate through deforestation and the draining of lakes and swamps, and the relationship between European imperialism and environmental degradation (of soil, water, trees, and air). Marsh rarely minces words and saves his harshest critique for the "crimes" of European imperialists who exported their ecological destruction to indigenous societies that had "interfere[d] comparatively little with the arrangements of nature."[77] Here is one of the more ominous passages in the book:

> The earth is fast becoming an unfit home for its noblest inhabitant, and another era of equal human crime and human improvidence, and of like duration with that through which traces of that crime and that improvidence extend, would reduce it to such a condition of impoverished productiveness, of shattered surface, of climatic excess, as to threaten the depravation, barbarism, and perhaps even extinction of the species.[78]

Indeed, Marsh's fiery tome can be read as an ecological critique of industrialization and economic growth. He takes issue with chemical and industrial waste, technologies that have accelerated environmental change, and above all the booming population of Europeans who see themselves as entitled to the world's land and resources. In short, Marsh, unlike Humboldt or Darwin, used the science of ecology to mount a campaign against industrial society and its effects. Marsh basically initiated the tradition of mobilizing ecology to undermine the stock narrative of progress that would undergird the industrial age. In striking contrast to most of his contemporaries, Marsh argued that humans had a moral duty to protect the planet, and his ethos of stewardship exerted great influence on the conservationists and preservationists who came to prominence around 1900.[79] Rachel Carson and Aldo Leopold took inspiration from *Man and Nature*, too, and Marsh is still relevant today.[80]

If G. P. Marsh is not a household name, John Muir probably is. Muir is remembered today as the founder of the Sierra Club (1892) and an ally of Teddy Roosevelt in the creation of the US national park system.[81] Born in Scotland and raised in Wisconsin, Muir traveled and hiked extensively throughout western North America before settling in rural California. He became an activist and an outspoken critic of Western expansion and industrial stresses on the natural environment. His activism led directly to the creation of Yosemite National Park and Sequoia National Park in California (both 1890). Muir was an ecological thinker who admired Darwin but departed from the anthropocentric views of most nineteenth-century ecologists. He held an organicist view of the natural world that stressed ecological interconnectedness and rejected the notion that humankind was "above" other species: "When we try to pick out anything by itself, we find it hitched to everything else in the universe."[82]

Muir was a master at rallying public opinion behind the preservation of wild places and wrote accessible articles in newspapers and magazines, such as *Harper's* and *Atlantic Monthly*. His writing is often a mixture of autobiography, romantic rumination, sobering scientific analysis, and social critique. In *Our National Parks* (1901), he made the case for expanding natural reserves, which would be used for hiking and leisure but not for resource exploitation, and emphasized the spiritual importance of wilderness.[83] In this passage, he blends a critique of "over-industry" with an exhortation to seek solace and regeneration in natural places:

> The tendency nowadays to wander in wildernesses is delightful to see. Thousands of tired, nerve-shaken, over-civilized people are beginning to find out that going to the mountains is going home; that wildness is a necessity; and that mountain parks and reservations are useful not only as fountains of timber and irrigating rivers, but as fountains of life. Awakening from the stupefying effects of the vice of over-industry and the deadly apathy of luxury, they are trying as best they can to mix and enrich their own little ongoings with those of Nature, and to get rid of rust and disease.[84]

Muir's activism eventually brought him into conflict with Gifford Pinchot, a one-time ally and hiking buddy who became the first chief of the US Forest Service in 1905. Although Muir and Pinchot are today remembered as two of the founders of conservationism, their views on the role of national parks and approaches to resource management were often strikingly at odds. Muir wanted national parks and reserves to be "preserved" from extensive resource exploitation; he did, however, approve of developing parks for leisure and recreation purposes, including the construction of hiking trails, roads, public facilities, and even the odd cabin or hotel. Placing himself in opposition to the narratives of industrial progress, Muir did not see forests as inert resources that should be sacrificed for economic expansion. Pinchot, by contrast, wanted natural areas to be "conserved" for sustainable resources consumption.[85] He was a student of Germanic sustained yield forestry, the theories of which he learned while studying at the forestry school in Nancy, and enjoyed the backing of much of the scientific community and the US federal government.[86] Some scholars have thus distinguished between Muir's "preservationism" and Pinchot's "conservationism" as divergent land protection strategies, and although this dichotomy might be too facile, the battle between Muir and Pinchot set the stage for the next 100 years' worth of debates over managing federally owned lands.[87] In recent years, the same conflict of values has arisen in debates over whether to drill for oil in the Arctic National Wildlife Refuge in Alaska.

The cause célèbre that brought this conflict into focus was an acrimonious dispute over a proposal to build a dam in the Hetch-Hetchy Valley, which was located inside Yosemite National Park. From the 1880s to the 1910s, Muir and Pinchot squared off on opposite sides of the debate. Pinchot supported the city of San Francisco, which sat 167 miles to the west of the Hetch-Hetchy Valley and which sought to dam up the valley for drinking water and hydroelectric energy.[88] Muir, who had helped to create Yosemite National Park, drew on the resources of the Sierra Club and his tremendous moral standing to counteract Pinchot's "ignorance." He argued that there were plenty of other, less ecologically significant and less protected valleys that could be dammed up. He also argued that Hetch-Hetchy could be developed as a tourist site rather than as a destructive energy installation.[89] Muir even appealed to Roosevelt directly to settle the dispute. In the end, though, Pinchot and the city of San Francisco won out, and the federal government passed the Raker Act of 1913 allowing the dam to be built.

The precedent set by the decision had ramifications that lasted well beyond the Progressive era. Historians argue that Hetch-Hetchy established Pinchot's conservationism, which was utilitarian and business oriented, as the chief approach to environmental policy for much of the twentieth century, in which forests were conceived as mere "timber" and wild places were seen as little more than exploitable "resources."[90] Yet even though Muir was on the losing side of this debate, his organicist ecology has remained since the 1890s a powerful vantage point from which to resist the ethos of industrial destruction and greed. Moreover, his impassioned activism and moral fortitude continue to inspire many in the sustainability movement today.

We are now 250 years into what sustainists would characterize as an experiment in unsustainable living called industrialism. In a sense, the Industrial Revolution has never drawn to a close: it has merely changed shape since the nineteenth century. Certainly, there are many continuities between the nineteenth century and our own. We still inhabit a stratified, urbanized society that runs on mechanization and fossil fuels and values economic growth above all else. The sites of heavy industry may have shifted in large part to different parts of the globe—from England, Belgium, and Germany to China, Mexico, Bangladesh, Indonesia, and so on—but the world is still in an industrial age. As we saw in this chapter, however, voices of caution and dissent have challenged industrial growth since the early nineteenth century. Scientists, political activists, and early ecologists laid the foundations for an alternative worldview that rejected the narrative of "progress," and if it is indeed true that "human activities in the industrial era" are the root cause of climate change and ecological destruction, then surely these early critics were warranted in their gloomy appraisal of industrialism. It would be wrong to argue that there was a self-conscious sustainability movement in the nineteenth century, but these thinkers and activists created the possibilities for the environmentalism and ecological economics of the mid-twentieth century, which in turn facilitated the emergence of the sustainability movement at the end of the century. In a very real sense, the cautionary voices of the nineteenth century still resonate in the present day.

Chapter 3

Eco-Warriors:
The Environmental Movement
and the Growth of Ecological
Wisdom, 1960s–1970s

One of the marks that distinguishes sustainability from classic environmentalism is the former's cheery optimism. Indeed, reading side by side the 2005 guidebook *Green Living*—a fairly typical how-to for sustainable living—with, say, Paul Ehrlich's doleful *Population Bomb* (1968) offers a case study in contrast. *Green Living* is constructive and buoyant whereas *Population Bomb* is frenzied and cynical.[1] Yet it's striking how much *Green Living* takes its inspiration not only from Ehrlich but from other titans of mid-century environmentalism—albeit with a noticeable shift in tone. Paul and Anne Ehrlich are cited approvingly in the opening pages of the book. The epigraph comes from the still-very-active David Suzuki. There are also references to the Leopold Center for Sustainable Agriculture, which, of course, is named after the esteemed Aldo Leopold. But gone is the gloomy tone, replaced instead by a heartening "You can do it!" attitude.

This brief observation illustrates how much the modern sustainability movement owes to the critics, intellectuals, and protestors of the 1960s and 1970s who raised awareness about environmental problems, advocated for social justice, and defended the rights of the oppressed. While the three Es of sustainability were rarely paired in the 1960s and 1970s, many of the basic concepts that shaped sustainability were clearly articulated before the 1980s. This chapter should not be taken as a comprehensive look at the environmental movement, about which there is reams more to say. Instead, it will examine in general terms some of the disparate sources that contributed to the holism of sustainability. Particular emphasis will be laid on the key ideas, associations, and scholars who developed the environmental movement and the success that environmentalists had in getting politicians, economists, and the public at large to think in ecological terms—a singular achievement that continues to inform the world of sustainability.

It is important to note that the reason that this book jumps from the late nineteenth century to the 1960s is not because the era of the two world wars has nothing to do with the history of sustainability. On the contrary, the privation and enforced self-sufficiency of the 1940s, in particular, is a worthy and relevant historical subject. However, the sustainists of the late twentieth and twenty-first centuries have tended to overlook this period in favor of the environmental movement that developed, in part, as a reaction to the hyperindustrialism, consumerism, and consumption that took root quickly in the postwar period. Thus this book focuses on environmentalism because of the explicit importance that this movement has played in shaping modern ideas about sustainability.

Perhaps the most significant achievement of the environmental movement in the 1960s was establishing "the environment" as a conceptual prism through which to view the world and humanity's place in it. The word "environment" dates back to the seventeenth century, but a specialized ecological connotation dates only to 1956; the term then entered common parlance in the 1960s. Thinking in terms of an all-encompassing, interconnected environment was an innovative act because it transcended the conceptual boundaries of "nature," which had been the ordinary way of reckoning air and water and dirt and animals since antiquity and which functioned as one half of an elementary binary that structured much of Western thought: the supposed opposition between nature ("out there") and human civilization ("in here").[2] The ecological sense of environment collapsed that binary and showed that humans and the built environment functioned as part of a broader ecosystem, defined as a community of living organisms and nonliving entities that interact in myriad ways and through which nutrients cycle and energy flows. No longer did it make sense to think of the natural world as solely "out there." Humans *are* the environment, both in the sense that we act as components of ecosystems and that ecosystems act upon us (via nutrients, pollutants, microbes, etc.). Donald Worster has traced this conceptual model—the "economy of nature"—back to the eighteenth century but acknowledges that ecological thinking gained widespread currency only in the 1960s.[3]

In this decade, a large and vocal community of scholars, student groups, activists, and environmental organizations drew on new (and often devastating) scientific research to raise awareness about environmental issues, press for social change, and advocate for governmental policies that favored the environment. The issues that received the most attention in this period included man-made pollution in air, water, and soil—mercury, pesticide chemicals, radioactive fallout, and so on—and the effects that pollutants had on human bodies and the ecosystem; the loss of wilderness and the need to preserve natural areas; the population explosion that stoked fears of widespread famine; and the depletion of natural resources (especially minerals and fossil fuels).[4] The hazards of mining and resource extraction (coal, uranium, gold, etc.) also received much attention.

The environmental movement rose to prominence because of a range of cultural phenomena that came together in the 1960s. First, the scientific study of pollution, especially radioactive fallout, smog, and chemicals, advanced greatly in this period, as did knowledge about the ability of pollution to affect humans and the ecosystem. Works by Rachel Carson and Barry Commoner, discussed below, translated this science to a horrified reading public. Second, a series of environmental disasters, such as oil spills in Cornwall (1967) and Santa Barbara Channel (1969), brought attention to the risks of reckless industrialism. Third, and more broadly, the period witnessed the rise of a culture of activism and the spread of countercultural liberal values, which challenged the status quo. In many ways, the environmental movement dovetailed with other activist movements in the 1960s: the opposition to the Vietnam War, the civil rights movement, and second-wave feminism, to name just a few.

In addition, the intensification of the nuclear arms race during the Cold War helped raise awareness about the fragility of life on Earth and eventually gave more credence to environmental science. When, in 1961, John F. Kennedy stood before the United Nations (UN) and warned that "every inhabitant of this planet must contemplate the day when this planet may no longer be inhabitable," he spoke of the threat of nuclear annihilation. But the message was clear that human technology and decision making could have a devastating impact on the environment and its ability to provide life-sustaining conditions for all living beings.

Many of the environmental issues discussed in the 1960s transcended the relatively narrow focus of the early conservationist movement, associated in the United States, as we saw in the last chapter, with Gifford Pinchot, Teddy Roosevelt, and John Muir, that revolved around property rights, land usage, the creation of national parks, and the appropriate use of natural resources on federally owned lands.[5] A more recent inspiration for the environmental movement was the work of a US Forest Service veteran and a professor at the University of Wisconsin named Aldo Leopold, who published his widely acclaimed *A Sand County Almanac* post-humously in 1949. Most of the book was a regional almanac but tucked in the back was an impassioned manifesto called "The Land Ethic" that laid out Leopold's philosophy about the potential for "harmony between men and land." Instead of approaching the natural world from the perspective of profit and self-interest, he argued that ethics and aesthetics should guide human interaction with the natural world: "In short, a land ethic changes the role of homo sapiens from conqueror of the land-community to plain member and citizen of it."[6]

Leopold became one of the most oft-cited environmental thinkers in the 1960s and 1970s and provided intellectual sustenance to new (and newly expanded) environmental groups that gained prominence at this time. The Canadian Wildlife Federation, today Canada's largest nongovernmental organization (NGO), opened in 1962. In Germany, numerous regional environmental groups appeared around this time and eventually banded together in 1975 to form the Bund für Umwelt und Naturschutz Deutschland. Of the ten major environmental groups that currently operate in the United States, four were founded between 1961 and 1970: the World Wildlife Fund for Nature (1961), the Environmental Defense Fund (1967), the Friends of the Earth (1969), and the Natural Resources Defense Council (1970). Satellite branches of these and similar organizations became very active on university campuses, too (meaning that environmental concerns now vied with more social and political ones in the world of student activism). The newer environmental groups in the United States joined six older, more established ones (including the venerable Sierra Club) to form a powerful lobbying, research, and advocacy bloc whose work would have a tangible effect on governmental policy, environmental awareness, and social practices.

This is not to say that the message of these scholars and organizations was welcomed with open arms. For many in the postwar generation, talk of the environment seemed to come out of no-where. The warnings of environmental scientists and activists appeared to run contrary to the relative stability and bounty of the postwar years. The mainstream environmental view at this time was that nature was essentially a cornucopia, that humans posed no serious threat to the natural world, and that there was the possibility for limitless growth.[7] As a result of entrenched attitudes toward the natural world, environmentalists spent a lot of time attempting to educate a largely apathetic public in basic environmental science and the need for behavioral change. The gains made by the environmental movement occurred only through advocacy campaigns at the grassroots level, printed polemics, and prolonged legal battles.[8]

Through easy-to-comprehend books geared toward a broad public audience and in the pages of the scientific journal *Science*, a series of conscientious scientists and academics brought the "environmental crisis," as it was often called, to the center of public consciousness and political debate. In the process, modern environmentalism took shape, distancing itself in many ways from an older, Teddy Roosevelt-style conservationism. This modern environmentalism was armed with cutting-edge scientific research, presented in a way that was meant to shock the reader and undermine naïve faith in the benign nature of technology and human practices. The classic environmental works of the 1960s are often disturbing, sensational, gloomy, foreboding, and alarming. The scientific findings seem to take the authors by surprise. Anger and anxiety leap off the page. The resentment is often palpable: How have humans been so *blind*, so *ignorant*, so *apathetic*? These works were not only critiques of, say, air pollution or overpopulation; they were also much deeper critiques of the harmful myths of technological progress, modernity, and the Industrial Revolution. The assumption was that Western industrialism had failed to live up to its own utopian promises and had instead created an ecological crisis that threatened life on

Earth. Environmentalists thus assumed the position of iconoclasts and bubble-bursters who served up sobering realities to a rapacious civilization hell-bent on growth and consumption at any cost. The knowledge and commitment of these authors greatly informed the sustainability movement of later decades, even as the gloomy tone and the narrow focus on the environment fell by the wayside.

The most influential of the classic environmental works from this period is Rachel Carson's 1962 *Silent Spring*, which ran as a serial in *The New Yorker* before appearing as a book. Carson had served as a marine biologist with the US Bureau of Fisheries before shifting in the 1950s to work full time as an independent science writer. *Silent Spring* became an instant and international best seller after its initial printing in the United States. (Today, Random House's Modern Library ranks it as the fifth most important nonfiction work of the twentieth century.) It was translated into numerous languages and made headlines in Sweden, Britain, the Netherlands, France, Italy, and Germany. The British House of Lords debated the book for five hours one day in 1963.[9] In the United States, *Silent Spring* set off a heated debate about the wisdom of using toxic synthetic chemicals to kill pests and increase agricultural productivity, but the book's influence extended well beyond those domains. Indeed, Carson's meticulously crafted book, which blends science with folk wisdom and polemical frankness, helped launch the environmental movement—although it did not create it, as many have suggested—and her impassioned admonitions contributed directly to new governmental policies in the Western world that banned or limited the use of many chemicals.

The book begins with a "Fable for Tomorrow" in which humans and animals begin to fall sick due to the excessive use of toxic chemicals. The birds eventually disappear and when spring rolls around, the skies are "silent" because songbird have all gone extinct. Carson then uses this dystopian image to initiate a scientifically rigorous, yet plainly readable, critique of the "lethal poisons" that have been inflicted upon the environment by industrial society: pesticides, insecticides, herbicides, fertilizers, and other chemicals, which she collectively renames "biocides." She states that, at the time of her writing, 500 new, often untested, chemicals hit the market every year in the United States. Carson casts environmental destruction as worthy of the same attention and concern as the prospects of apocalyptic nuclear holocaust: "Along with the possibility of the extinction of mankind by nuclear war, the central problem of our age has therefore become the contamination of man's total environment with such substances of incredible potential for harm."[10] Perhaps her most chilling message was that these chemicals were not isolated "out there" but flowed through the ecosystem into the bodies of mammals, including homo sapiens, and other vertebrate animals. Some were linked to cancer in humans. She also revealed that overloading the environment with insecticides was both useless, since insects adapted to their effects, and harmful, since the chemicals had delayed effects in the ecosystem. She saved her most vocal criticism for a widely used insecticide called dichlorodiphenyltrichloroethane (DDT) that entered human organs and tissue via beef from cows fed on DDT-laden hay. DDT biomagnification— increased concentration as one moves higher up the food chain— also affected birds who consumed plants, insects, and fish with high DDT levels, sterilizing the birds or causing the thinning of egg shells that prevented maturation. DDT had a devastating impact on the population of bald eagles, the symbol of American freedom, and many other increasingly "silent" bird species, as Carson persuasively demonstrated.

The reaction to Carson's bombshell says a lot about the public reception of environmental science in the early 1960s. Carson was denounced as a communist, an amateur, a hysterical woman, and a troglodyte who wanted to reverse the hard-earned gains of industrial modernity. Those were vicious lies, but Carson had succeeded in questioning the "modern way of life" that saw new technologies as risk free and put blind faith in the work of scientific experts. The book became a polarizing cause célèbre. It helped define environmentalism as a new kind of worldview that rejected ecosystem destruction and industrial growth in the name of "progress," and it forced citizens in industrialized countries to take sides on environmental regulation.

Monsato, Velsicol Chemical Company, DuPont, American Cyanamid Company, and other giants in the chemical industry, with their entrenched interests, worked hard to undercut Carson's credibility and defend "harmless" insecticides such as DDT. But others were more receptive to her message. In fact, the book had galvanized enough support and interest that President Kennedy in 1963 asked his Science Advisory Committee to investigate the research.[11] Carson was called to testify at committee hearings. Surprisingly, the committee vindicated Carson entirely and confirmed the destructive effects of biocides. (Incidentally, Carson died of breast cancer in 1964 at the height of the controversy.) As a result, the federal government began to regulate the chemical industry, and DDT was eventually banned between 1970 and 1972. By 1975, all of the chemicals discussed in *Silent Spring* had either been banned or severely restricted in the United States, and countries in other parts of the Western world adopted similar policies in the 1970s. Within a few years, chemical levels dropped in the environment, and many bird species began to make a recovery. In a broader sense, though, Carson raised new consciousness about the ability of humans to degrade the natural world and triggered a far-reaching international dialogue about the role of environmental regulation.

Stewart L. Udall's *The Quiet Crisis* of 1963 built on Carson's successes. Udall was a very progressive member of the US federal government and served as Secretary of the Interior for nearly all of the 1960s. Over the course of his impressive career, he oversaw the creation of thousands of square miles of national parks and other protected areas and played a crucial role in the promulgation of groundbreaking environmental laws. His book of 1963 echoed Carson's by warning about the risks of air, water, and soil pollution. He also wrote matter-of-factly and authoritatively about resource depletion and the decline of wilderness areas in the United States. Perhaps most important, Udall provided the fledgling environmental movement with its first bona fide history, situating environmental activism within an esteemed tradition that dated back at least to John Muir.[12]

What was so fascinating about Udall's book was that it was written by a governmental insider and one with tremendous moral rectitude, who was very openly in favor of the environmental movement. Indeed, Udall would have ample opportunities over the course of the decade to put his money where his mouth was in dealing with what other bureaucrats were afraid to call a "crisis." The presence of someone like Udall in the US federal government was a major coup for the environmental movement—as was his ability to float above the fray of political partisanship and his willingness to engage and inform the public with, for instance, his 1963 best seller.

Joining Udall on the best seller rack was a decidedly less mild-mannered and much more polemical Stanford biologist by the name of Paul R. Ehrlich. In his aforementioned book, the 1968 *The Population Bomb*, Ehrlich (and his uncredited wife, Anne) sounded the alarm bells on global overpopulation and the imminent threat of famine, war, starvation, and catastrophic environmental destruction. Ehrlich borrows a page from Rachel Carson by beginning his analysis with a story about a dystopian future: an Earth crammed with hundreds of billions of hapless humans. He then provides hair-raising warnings about what would happen to much of humanity if the world's population continued to double every 35 years. Here's a fairly representative passage: "Of these poor, a *minimum* of ten million people, most of them children, will starve to death during each year of the 1970s. But this is a mere handful compared to the numbers that will be starving before the end of the century."[13] Ehrlich also borrowed from Carson by denouncing the naïve faith in science and "technological fixes" to solve problems that had, in large part, been generated by both science and technology.

Even if Ehrlich's more dire warnings about "mass starvation" never quite materialized, he did succeed in separating population studies, which he saw as an ecological discipline, from the drab study of demographics, even as some of his more extreme suggestions for population control nauseated many readers.[14] He also raised quite a lot of awareness about the threat of an overpopulated world. He inspired the foundation of a group called Zero Population Growth, which counted over 30,000 members in 1970 and advocated for stabilizing world population at a sustainable level. He even earned an audience with President Nixon via the 1970 Commission on Population Growth and the American Future. Ehrlich and John P. Holdren advised the committee, putting forth a number of suggestions about how best to curb population growth (sex education, contraception, abortion, etc.). It was in the context of this research, along with an ongoing exchange with his rival Barry Commoner, that Ehrlich (and Holdren) developed a hugely influential formula to quantify and measure human assaults on the environment.[15] The formula $I = PAT$, as it was eventually known, meant that one could understand environmental "impact" by multiplying together the "population," "affluence," and "technology" of a society. The equation remains important in environmental science, if less frequently used.

Ehrlich and Holdren's article on quantifying humans' environmental impact appeared in *Science*, the Washington, DC-based journal with a sterling reputation whose articles were (and still are) written in an intelligent but highly readable magazine style. Indeed, *Science* became an important outlet for a new breed of intellectual in this period: the scientist-activist, who wielded privileged technical knowledge to wage public campaigns for environmental justice. That is, many scientists in the late 1960s and 1970s had a "message" that transcended the neutral tones of the academic researcher. David Suzuki became the preeminent model of this crusading scientist. Born in Canada in 1936, Suzuki studied biology and zoology in the United States before returning to Canada where he not only taught university-level science but also hosted a series of successful radio and television programs aimed at raising environmental consciousness. He was, moreover, an energetic activist who put his reputation behind issues ranging from logging practices to climate change, and his eponymous foundation has been a force in the environmental world since 1990. Ehrlich, likewise, marshaled his population analyses into a call for public action: the latter chapters of *The Population Bomb* provide a blueprint for citizens to pressure authorities into adopting policies of stricter population control.

Science also published two other critical articles in the 1960s that made a lasting impression on the environmental movement. The first was a 1967 article by Lynne White Jr., a professor of medieval European history at the University of California at Los Angeles, called "The Historical Roots of Our Ecological Crisis." In this pathbreaking and much-debated work of environmental history, White historicizes the ecological crisis of the 1960s by arguing that Christianity's deep-rooted hostility toward the natural world facilitated the ecological devastation of the Industrial Revolution. He contends that the conventional reading of Scripture in and after the Middle Ages was that human beings possessed a rightful dominion over nature, that human beings were the most important beings on Earth (anthropocentrism), and that beings other than humans were necessarily soulless and inferior to humankind. Echoing many of the environmentalists of the day, White works this historical critique into a repudiation of destructive modern technologies, which he sees as the embodiment of hostile attitudes toward nature: "Our science and technology have grown out of Christian attitudes towards man's relation to nature which are almost universally held not only by Christians and neo-Christians but also by those who fondly regard themselves as post-Christians."[16] White, in effect, triggered a broad debate (that lasts to the present day) about not only the relationship between religion and environmental degradation but also the myriad cultural factors that have contributed to conceptions and uses of nature.

The second article was Garrett Hardin's 1968 "The Tragedy of the Commons." In that famous work, Hardin, an ecologist at the University of California at Santa Barbara, sounded off on the inability of technology to solve humanity's ills, among which he listed overpopulation and nuclear warfare. His most important move, however, was to put an environmental twist on a longstanding critique of Adam Smith's contention that private self-interest leads ultimately and uniquely to public benefit. To make his case, Hardin draws on a hypothetical example of a commonly held pasture that is overexploited by a group of local herdsmen, each of whom has the desire to maximize self-interest by adding more livestock to the fold. In the end, though, the pasture is overgrazed and the well-being of *all* the herdsmen is threatened—and therein lies the tragedy. As Hardin put it, "freedom in a commons brings ruin to all."[17]

Some took Hardin's article as a call for increased privatization of natural capital, but most understood it as a warning against greed, ignorance, and apathy. Certainly the commons (rivers, oceans, air, etc.) has suffered greatly, as have those who rely on its essential life-supporting ecosystem services (i.e., everyone). Later in life, Hardin formulated three laws of ecology, which remain influential: (i) We can never do merely one thing; (ii) there is no away to throw to; and

(iii) the formula I = PAT, mentioned above, as a way to understand humankind's impact on the environment.

Perhaps the most emblematic of the scientist-activists of the classic environmentalist era was the indefatigable Barry Commoner. Commoner, who passed away in 2012, was a Harvard-trained zoologist who worked for many decades at Washington University in St. Louis, where he helped to develop the science of ecology. (He also ran for president, quite unsuccessfully, on a third-party ticket in the 1980 election.) He became well known in the 1950s and 1960s for his outspoken opposition to nuclear weapons testing and helped to produce powerful research that demonstrated the devastating ecological effects of nuclear fallout. On a deeper level, though, he became a major critic of the way in which richer countries extended the colonial project, in a sense, by exploiting the resources and human labor of poorer countries. His main contention was that wealthier countries had an obligation to help impoverished ones reach a level of economic well-being that would prompt self-imposed family planning and ultimately stabilize the global population.

But Commoner's most famous work is his 1971 best seller, *The Closing Circle: Nature, Man and Technology*, which was written in response to the muddled debates about environmental degradation set off by the first Earth Day in 1970. In this terrifying tour de force, Commoner describes in the bluntest of terms how man-made technologies have devastated the "ecosphere"—Earth's life-supporting spheres. Along the way, he profiles the tragically filthy air of Los Angeles (largely from automobile lead emissions), the mercury pollution that comes from chlorine production, and the inability of the environment to break down man-made synthetics, such as those found in laundry detergents. His stress on the "technological flaw" of human society led some critics to argue that Commoner overlooked other factors that had contributed to the ecological crisis.[18]

But that misgiving did not blunt the impact of this now-classic work of ecology. Indeed, the four "laws of ecology" that Commoner enumerated in the book became his most lasting contribution to the environmental studies:

1. Everything is connected to everything else.
2. Everything must go somewhere.
3. Nature knows best.
4. There is no such thing as a free lunch.[19]

These very useful and pithy laws, upon which Commoner expands in the pages of the book, provided the fledgling science of ecology with shape and character, and one can still see these fertile aphorisms behind much of the innovations of the sustainability movement, from biomimicry and biodegradable products to renewable energy and closed-loop industrial production.

Two more learned critics are worth mentioning: Jane Jacobs and Henri Lefebvre. Although neither is considered an environmentalist in the ordinary sense of the word, both were influential writers and activists who had an immeasurable impact on urban ecology, urban theory, and urban planning. In the 1960s and 1970s, both Jacobs and Lefebvre influenced social activists in North America and Western Europe, but since the 1980s, they have been more closely associated with the idea of the modern sustainable city: a livable, walkable, safe, democratic, beautiful, and ecologically sound conurbation. The "environmental" content in the works of Jacobs and Lefebvre is, indeed, fairly light, in the sense that neither theorist was particularly concerned with things like pollution, energy sources, waste, or natural resources, and both tended to focus on the "social" and "economic" aspects of urban life. However, the sustainability movement beginning in the 1980s adopted and modified the works of Jacobs and Lefebvre and integrated environmental components into their respective socio-economic arguments and insights. Today, Jacobs is considered the godmother of New Urbanism—perhaps the world's most important urban design movement since the 1990s—and Lefebvre not only helped inspire the May 1968 protests in France but remains a cherished figure in left-wing urbanism.

Born in 1916, Jacobs spent the first half of her life as a writer and community organizer in New York City before moving to and settling in Toronto. She was instrumental in preventing the construction of a highway in Lower Manhattan and an expressway in downtown Toronto. Her most famous work is the classic 1961 *The Death and Life of Great American Cities*, which is a biting and trenchant "attack on current city planning and rebuilding."[20] Jacobs excoriates Ebenezer Howard's artificial Garden City and Le Corbusier's ugly and utilitarian Radiant City. She also takes aim at ill-conceived suburbanization plans ("the Great Blight of Dullness") and seemingly nihilistic "urban renewal" efforts that knock down perfectly good infrastructure and inhibit community-building. She saved her fiercest critique, though, for Robert Moses, the Baron Haussmann-like figure who bulldozed the old city and reoriented New York around cars, freeways, and parking lots.

Jacobs is at her best when she defends pedestrian-oriented public spaces, cultural and architectural diversity, and the importance of relatively dense, "mixed-use" neighborhoods (commercial, residential, recreational). Jacobs absolutely adored the pageantry and regulating functions of the sidewalk. She also analyzes, with refreshing frankness, the de facto social and ethnic segregation that beset many American cities in this period—due in large part to the racism and classism of moneylenders and municipal authorities— and offers excellent ideas about how to go about "unslumming" the ghetto. Her defense of mixed-use areas and her emphasis on safe and walkable communities has directly influenced the idea of the wisely planned and socially just sustainable city.

Henri Lefebvre was a French Marxist, academic, and urban theorist who condemned the capitalistic privatization of European cities and defended the idea that cities belong to the people who dwell inside them. After penning numerous works on sociology and philosophy, Lefebvre moved to the quite radical University of Nanterre in 1965, where he ran the Institut de Sociologie Urbaine. His writings and lectures helped inspire the student uprising in France in May of 1968, which began as a protest against conditions within institutions of higher learning, but when workers joined the students—11 million workers suddenly went on strike—the protests expanded into a wide-ranging critique of French society and the administration of President Charles de Gaulle. Himself inspired by the popular takeover of Paris—or many parts of it— Lefebvre turned full time to the study of urbanism. His subsequent works include *Le droit à la ville* (The right to the city), published late in 1968, and the 1974 *La production de l'espace* (The production of space).

Three main themes emerge from Lefebvre's rather philosophical and Marxian works. First, the people (and not the bourgeoisie) have a "right to urban life" and should construct a city that is public and equitable. Second, cities have "use value" (inherent value as entities that are used) that transcends the commodified "exchange value" that has been imposed on them by the capitalist classes. Third, the use of space in modern cities does not take shape haphazardly but is rather "produced" via a particular mode of production; cities are now devised to favor the wealthier classes who are able to privatize and appropriate urban space.[21] Today, Lefebvre's steadfast defense of the publicly owned and egalitarian city remains a source of inspiration for everyone from German anarchists to middle-of-the-road urban planners.

The tactics, research, and legislative successes of the classic environmental movement of the 1960s and 1970s created the conditions of possibility for the philosophy of sustainability. I purposely avoid the word "caused" since sustainability is not an inevitable outcome of the environmental movement, but at the same time *it could not have existed without it.* Environmentalism facilitated the integrated systems thinking of sustainability that truly began to take shape in the 1980s.

What were those successes? In addition to raising public consciousness about the world's mounting ecological crisis, the environmental movement by the early 1970s could boast of a number of important gains. This is not to imply that environmentalists got everything that they wanted, and the near-consensus among environmentalists was that the world remained too polluted, too overpopulated, too wasteful and consumerist, and too socially unjust. But the movement's faithful could point to the political arena for evidence of obvious achievements. In the United Kingdom, the infamous great smog of 1952 in London prompted the pioneering Clean Air Act of 1956. In the United States, a rapid-fire series of federal acts in the 1960s ushered in an unprecedented level of environmental regulation and protection. Stewart Udall, in fact, played midwife to many of these groundbreaking environmental laws: the 1963 Clean Air Act (amended in 1970) that regulates air pollution at the national level; the Wilderness Act of 1964 that protects millions of acres of federally owned "wilderness", the Land and Water Conservation Fund Act of 1965, the Solid Waste Disposal Act of 1965, the Endangered Species Preservation Act of 1966, the Wild Scenic Rivers Act of 1968, the National Environmental Policy Act of 1970 that requires environmental impact statements for federal projects and actions, and the Clean Water Act of 1972 that regulates toxins and other water pollutants. The culmination of these efforts came in the form of the crucially vital US Environmental

Protection Agency, which was created by President Nixon in 1970 and whose mission remains to "protect human health and the environment." Similar legislative acts and agencies appeared in other Western countries between the 1970s and 1990s.

Sticking with the political arena, environmentalists in numerous countries formed a new, internationally oriented Green Party (some of which grew out of "ecology parties"). By the mid-1970s, the Greens had functional branches in Belgium, England, and Germany. In fact, Belgium was probably the first country to elect a Green Party candidate to a parliamentary seat, in the Belgian general election of 1981. By the mid-1980s and 1990s, Green parties had popped up all over Europe and North America, and the German Green Party in this period became a moderately influential bloc in the West German Bundestag. The advent and growth of these parties meant that, for the first time in many parts of the world, there was a party that was explicitly dedicated to the defense of "green politics": ecological wisdom, grassroots democracy, nonviolence, and social justice. In the United States, the consumer advocate and ecological activist Ralph Nader brought attention to the party with a series of high-profile presidential campaigns.

Perhaps even more important than the emergence of the global Green Party was the foundation of international institutions and conferences devoted to addressing the world's presumptive ecological crisis. The heavyweight, in this domain, is the UN Environment Programme (UNEP), which has coordinated international environmental efforts since 1972 (and which today is one of the leading NGOs that promotes sustainability). UNEP was born out of the UN Conference on the Human Environment held in Stockholm in June of 1972—although the UN General Assembly had convened the conference in 1969. The Stockholm conference was the first in a long series of UN-backed meetings that advanced international environmental law and protection and which, by the late 1980s, had become the main force and outlet of expression for the adherents of the sustainability movement. The 1972 conference spawned new research in environmental studies and produced a fairly blunt "Declaration" involving 7 proclamations and 26 principles that address issues of environment and development.[22] Since Stockholm, the idea of collaboratively addressing environmental problems has been a cornerstone of international relations and accords.

Finally, environmentalists could point to the creation of Earth Day and the rediscovery of recycling as positive results of the movement. Earth Day (or Earth Week, in some places) was first championed by the peace activist John McConnell and later by US senator Gaylord Nelson. The date was fixed as April 22 in most places, the vernal equinox in others, and was celebrated for the first time in 1970. It was a huge international success, and the UN adopted Earth Day as an annual, global celebration in 1971. Since then, Earth Day has become an international holiday of sorts that is meant to promote "environmental education" and "public policy solutions" to environmental problems.[23] Since the early 1970s, Earth Day has taken as its symbol the now-taken-for-granted photograph of the Earth furnished by NASA astronauts that seems to confirm via pictorial gravitas the obvious truth that our little "blue marble" is a relatively small and fragile ecological unit. It's all we have.

In terms of recycling, which has been called the greatest success of the environmental movement, the 1970s witnessed a rediscovery of the practice, which had been a standard feature of the war economy during the 1940s and which, in various ways, had been practiced since time immemorial in pre-industrial societies. The Romans recycled bronze coins into statues, and all world societies that possessed metals tended to reuse them, given their value. Paper recycling existed in feudal Japan and in early America. The Mayans, Egyptians, and Europeans reused building stones whenever possible. Napoleon Bonaparte's Arc de Triomphe du Carrousel (finished in 1808), next to the Louvre, reused pink marble columns that had been salvaged from the burnt-out ruins of the Château de Meudon. During World War II, military needs meant that goods such as nylon, rubber, and scrap metal were rationed and recycled. However, the relative plenty of the postwar years pushed recycling to the backburner, as high-consumption lifestyles formed the basis of a new Western cultural identity. But in the 1970s, there was renewed interest in (and new technologies for) the recycling of glass, paper, and metals. Recycling is now commonplace in the Western world and, in fact, is required in many large cities. Recent research has confirmed that recycling reduces the overall throughput and carbon footprint of our consumer society.[24]

What about the social movements of the 1960s? Did they, too, help create the conditions that made the sustainability movement possible? Yes, but perhaps in a less obvious way. It is easy to overlook, but the third E of sustainability is equality or social justice, and even though the activists of the 1960s rarely paired together social issues with environmental concerns, their unswerving support for equal rights, peace, and human safety are certainly part of the historical background of the sustainability movement. Although Martin Luther King, Jr. occasionally criticized greed and materialism, it would probably be a stretch to call him an "environmental" thinker. The same goes for many of the other figures and groups that played a role in the concurrent anti-war movement, the women's movement, and the civil rights movement. One can search quite a long time in the records of the Students for a Democratic Society or in the speeches of Malcolm X before coming across content that could be understood in terms of environmental justice, although some New Left groups connected the dots between pollution and profit.[25] (However, some anti-war figures, such as Abbie Hoffman, later metamorphosed into environmental activists.)

Likewise, environmental concerns seldom took center stage in the mass demonstrations of the 1960s. But these observations are decidedly unfair since social activists had other axes to grind: the establishment of equal rights, the end to the Vietnam War, and the amplification of people power. One could say that social and environmental activists took different approaches to what were beginning to be called quality-of-life issues.

Perhaps one exception was the anti-nuclear movement, which in many ways spoke to both "social" and "environmental" concerns. The most prominent group opposed to nuclear testing and nuclear power in North America was Greenpeace, which took shape between 1969 and 1972 in Vancouver and which was originally an organization devoted to disrupting nuclear tests through aggressive direct actions. The early members of Greenpeace saw themselves as activists that defended social well-being by ensuring the preservation of a safe and unpolluted environment. By the same token, the Green Party in Germany grew out of a concerted opposition to nuclear power and nuclear weapons, as did environmental parties in other parts of the world.

In many ways, then, the environmental and social movements of the 1960s and early 1970s are the direct predecessors of and inspiration for the modern sustainability movement. That lineage is marked by a number of current practices, concepts, and concerns with strong roots in that period of fruitful activism. First, we have the elaboration of an ecological conception of "the environment" that assumes a complex relationship between humans, human technology, and the built environment, on the one hand, and ecosystem use and degradation, on the other. It is this sort of integrated systems thinking that provides the essential background to the idea of sustainability. Ecology, in a sense, created a new kind of worldview that allowed for, more critical perspectives on the romantic narrative of the "progress" of industrial society—a narrative that was either apathetic toward or overtly hostile to the natural world.

Second, and quite related, is the critique of technology articulated by nearly all of the environmentalists discussed above. As discussed in the previous chapter, technophobia and technoskepticism go back at least to the Luddites and other machinebreakers, but environmentalists narrowed in on the capacity of modern technology to destroy the environment: air pollution from burned fossil fuels and industrial emissions, the toxins from synthetic chemicals that harmed animals and depleted soils, the radiation from nuclear waste, industrial practices that polluted the water table, and the unquestioning acceptance of new technologies that privileged the big, the brown, and the centralized.

Third, the social activists who pressed for peace, equality, and democratic rights remain inspirational to those in the present day who fight for social justice, even if awareness about food safety, the right to live in a clean environment, and the links between environmental degradation and armed conflict, for instance, was still some time off. Fourth and finally, the 1960s essentially spawned the scientist-activist who was aware of environmental problems and was not shy about rallying the public to do something about it. Many, although certainly not all, of the social, economic, and environmental problems with which the sustainability movement concerns itself in the twenty-first century were brought to light by scientists, economists, and social activists in the 1960s and 1970s. What was so revolutionary about the sustainability movement in the 1980s and 1990s, as we shall see in the following chapters, is that it aggregated a set of evidently interconnected problems—problems that happened to have social, environmental, and economic *dimensions*.

Chapter 4

Eco-Nomics

"La civilisation du toujours plus"

—Bertrand de Jouvenel

The environmental movement of the 1960s and 1970s overshadows a second, less heralded intellectual development that took place at the exact same time: the birth of "ecological economics." A cluster of nonconforming economists in this period drew on the fledgling science of ecology to rethink many of the assumptions of neoclassical economics, with its "growthmania," general indifference toward pollution and ecosystem destruction, and dogmatic belief that "tastes and preferences" are innate in humans rather than culturally shaped. What emerged was a new school of thought that integrated ecological concerns into an essentially capitalist economic framework.

These iconoclasts brought together the dual nature of the Greek word "oikos" (literally: household), which is the etymological root of both "economics" and "ecology." They asserted that the human "household" could not exist without a healthy and functional natural environment. This has become the essential insight of economic sustainability—the second "E" of sustainability: that the world needs economic systems that exist harmoniously with nature (and which promote social equality and justice). Those who practice the economics of sustainability in the present day— William E. Rees, Mathis Wackernagel, Peter Victor, Tim Jackson, Richard Heinberg, and many others—are the heirs of these early critics who challenged the hegemony of business-as-usual economics.

First-wave ecological economics shares the readability of the classic environmental works discussed in the previous chapter. The main authors associated with ecological economics—E. J. Mishan, E. F. Schumacher, Kenneth Boulding, Howard T. Odum, Nicholas Georgescu-Roegen, Herman Daly, Amory Lovins, and the members of the shadowy-sounding Club of Rome—went out of their way to write nontechnical books that were meant to appeal to the average-educated reader. Collectively, these authors ask deep and penetrating philosophical questions: What is the point of endless economic growth? What are the environmental costs of a wasteful and fossil-fuel-addicted consumer society? What is the best way to measure the well-being of a society? What is the role of economics in ensuring that human society remains within its ecological limits and avoids overshoot and collapse? How can nature, society, and the economy be studied as a single system? Is modern technology more harmful than beneficial?[1] The theoretical (and, at times, utopian) approach means that many of these books read more like "political economy" than economics—political economy being the older, speculative cousin of economics with deep roots in moral philosophy.[2] Either way, all of these economists operate within a capitalist structure, meaning that they assume the continued existence of private property, the private ownership of the means of production, distribution, and the exchange of wealth, a marketplace of goods and services, monetary currencies, and so on.

However, the ecological economists cautioned quite strongly against the toxic cocktail of overpopulation, overconsumption, pollution, and resource depletion, and in this way they were as skeptical about the promises of industrial modernity as the environmentalists who informed them. They also tended to favor a more regulated economy in which the state played an active role in imposing environmental taxes and regulations. To be clear, they rejected communism and conservative corporatist economics as well as laissez-faire capitalism (and the nightwatchman state) on the grounds that all of these economic systems had proven equally powerless in preventing ecosystem destruction.

The main object of criticism, however, was not a command economy but neoclassical economics, symbolized most powerfully in the United States by Milton Friedman and the Chicago School of Economics.[3] Indeed, Friedman brought together most of the beliefs that the ecological economists found objectionable: the identification of laissez-faire economics with "freedom"; monetarism, in which central banks control the money supply and influence output and price levels with economic growth in mind; the faith placed in pricing and "supply and demand" to regulate shortages and surpluses (of natural resources, for instance); the unshakable conviction that the purpose of an economy is perpetual growth; the belief in the benign nature of modern technologies; the complete apathy toward pollution and environmental degradation; and, finally, the notion that one could view a polluted environment as something separate from a healthy economy (later called "decoupling"). The economists profiled here also had serious misgivings about the theories of other economic heavyweights, including John Maynard Keynes and Friedrich Hayek, who "exogenized" (i.e., ignored) the environment and supported the idea that economic growth is generally a positive thing.[4]

The first wave of ecological economists met with moderate success. E. F. Schumacher was invited to the White House by President Carter in 1977 to discuss his much-discussed collection of essays, *Small Is Beautiful*. Herman Daly found unlikely success at the World Bank, and helped make the financial juggernaut "more environmentally sensitive and literate."[5] Amory Lovins founded the hugely successful Rocky Mountain Institute, which remains a leader in sustainability research and consulting work. The Club of Rome, a think tank founded in 1968 by an Italian industrialist and visionary named Aurelio Peccei, published the best-selling *The Limits to Growth* (authored by Donella H. Meadows, Dennis L. Meadows, Jørgen Randers, and William W. Behrens III) in 1972, which triggered a wide-ranging debate that lasts to the present day about the wisdom of promoting constant growth. Yet, on balance, the heretical critiques of these economists and systems theorists did relatively little to shake the dogmatic faith in a growth-based, deregulated economy. Relatively few universities, banks, or governments employed the disciples of these thinkers between the 1960s and 1990s. Only in recent years, with mounting resource pressures and the public acceptance of anthropogenic climate change, have ecological economists come to gain more prominence and credence vis-à-vis the mainstream.

What follows is a thematic overview of the core features of the first wave of ecological economics. The analysis is thematic, rather than centered around each author, since the same critiques, concerns, and proposals appeared in numerous works of the late 1960s and 1970s. While there was a fair bit of disagreement between these economists, much of which won't be discussed in this chapter, there was nonetheless a general consensus about the deficiencies of neoclassical economics, which guided economic policy throughout much of the Western world and beyond. Indeed, one of the features that distinguish first-wave ecological economics from the more recent economics of sustainability is the emphasis on condemning mainstream economics rather than focusing on how to construct a "green economy." One notable exception is Amory Lovins's pioneering work of 1977, *Soft Energy Paths: Toward a Durable Peace*, which laid out a blueprint for weaning industrial societies from "hard energy" (nuclear, fossil fuels) and getting them to run on safe and renewable energy sources.[6] But to read the classic ecological economists is to be exposed to deep-seated reservations about the faults of an economic system that led, in the minds of these economists, to the ecological crisis exposed by scientists at mid-century. The ecological economists were as disillusioned with industrial consumer society as the environmentalists of the same era, but they expressed their disillusionment primarily in economic terms. They believed that the Industrial Revolution did not need to be a *permanent* revolution and that society needed a stable, just, and ecologically sound economy—heresy, indeed, for the capitalist societies of the Cold War era.

THE COSTS OF GROWTH

The first and perhaps most important critique of these economists centers on the rigid belief that a "healthy" capitalist society requires perpetual economic growth—that is, an unending increase in aggregate throughput, consumption, and/or output. As Robert A. Nisbet noted decades ago, "growth" is a very powerful metaphor that tends to be associated with positive attributes in Western societies.[7] To oppose growth is somehow to oppose the "development" or "evolution" of society and favor "stagnation" or even economic recession. As Herman Daly put it in his 1977 *Steady-State Economics*,

"The verb 'to grow' has become so overladen with positive value connotations that we have forgotten its first literal dictionary denotation, namely, 'to spring up and develop to maturity.' Thus the very notion of growth includes some concept of maturity or sufficiency, beyond which point physical accumulation gives way to physical maintenance."[8] Even Adam Smith and Keynes had suggested that a society could, in theory, reach a level of wealth at which point growth would take a back seat to human happiness.[9] Yet to question the benefits of growth in the 1960s was tantamount to telling a sixteenth-century pope that the Earth orbited the sun.[10]

What do these economists see as the problem with growth? What troubled them so much was that industrial society seemed to have enough stuff, enough wealth, and enough people. Anything more just seemed excessive—and the desire for "more" appeared pointless and ideological. Mishan, Daly, and Schumacher all offer pointed critiques of the orthodox economic belief that a "mature" economy must grow by 3% to 5% per year to be considered strong and viable. In *Small Is Beautiful*, Schumacher makes the case that the ultimate consequences, aims, and value of growth go unquestioned by economists:

> For example, having established by his purely quantitative methods that the GNP of a country has risen by, say, five percent, the economist-turned-econometrician is unwilling, and generally unable, to face the question of whether this is to be taken as a good thing or a bad thing. He would lose all his certainties if he even entertained such a question: Growth of GNP must be a good thing, irrespective of what has grown and who, if anyone, has benefited. The idea that there could be pathological, unhealthy growth, disruptive or destructive growth is to him a perverse idea which must not be allowed to surface.[11]

Not only is the quest for "more" pointless and arbitrary, these economists argued, but also it engendered devastating ecological (and social) consequences. In his 1967 *The Cost of Economic Growth*, Mishan derides "the iron clutch of economic dogma," which he associates with the antiquated economic thought of the nineteenth century, and analyzes the environmental "costs" that come with a consumer society overloaded with new-fangled "anti-drudge devices."[12] It is here that we see why Mishan and others were later called "ecological economists." Mishan provides a partial list of the effects of "unchecked commercialism" in the industrialized world:

> The erosion of the countryside, the "uglification" of coastal towns, the pollution of the air and of rivers with chemical wastes, the accumulation of thick oils on our coastal waters, the sewage poisoning our beaches, the destruction of wild life by indiscriminate use of pesticides, the change-over from animal farming to animal factories, and, visible to all who have eyes to see, a rich heritage of natural beauty being wantonly and systematically destroyed.[13]

Mishan was not alone in his attempt to integrate ecological factors into the field of economics—something that neoclassical economists rarely bothered to do. Daly and the Club of Rome, for instance, each argued that advanced economies needed to figure out a way to live within their "biophysical limits," which meant, for them, a necessary reduction in throughput in the economy, a decrease in pollution levels, a stabilization of the population, a slower rate of resource consumption, and an increase in renewable sources of energy. Indeed, one's attitude toward "limits" became the litmus test in this period for continued association with mainstream economic thought. In the debate on growth that raged in the 1970s, the main division between the ecological and neoclassical economists centered on whether an industrial economy could grow indefinitely.[14]

The ecological economists were adamant that it could not—hence the Club of Rome's title: *The Limits to Growth*. The original edition of 1972 reported:

> Global ecological constraints (related to resource use and emissions) would have significant influence on global developments in the twenty-first century. *Limits to Growth* warned that humanity might have to divert much capital and manpower to battle these constraints—possibly so much that the average quality of life would decline sometime during the twenty-first century.[15]

The Club used cutting-edge computer models and systems theory to demonstrate that "exponential growth" in "population, food production, industrialization, pollution, and consumption of non-renewable resources" was unsustainable, and that a series of "feedback loops" served to exacerbate the consequences of growth.[16] That is, the growth that the West had experienced since the Industrial Revolution could be considered, in many cases, a kind of backfiring "uneconomic growth"—a term that Daly later coined to suggest that some forms of growth impair not only future growth but also future quality of life.[17] The ecological lesson is clear: growth of this sort and at this rate cannot last forever. In his book of 1977, Daly argues relentlessly that the obsession with mathematics and equations in economics had drowned out questions of ethics and values. As a result of this bigger-is-better worldview, economists had chosen to ignore the ecological and social consequences of growth. Yet "the biophysical facts have asserted themselves in the form of increasing ecological scarcity: depletion, pollution, and ecological disruption."[18]

Isn't growth "natural" and "necessary"? The ecological economists argue that it is not. Mishan, Daly, and Schumacher, in their
respective works, patiently wade through the pro-growth arguments and respond to them in a point-by-point refutation. For
them, rampant growth is the by-product of a concerted effort on
the part of Western governments since World War II to pump-
prime the economy with government spending, interest rate alteration, and tax reductions meant to stimulate growth. That is not to
say that these economists are against growth of any kind; few of
them would oppose, for instance, an exercise studio that decides to
open a second location and thus "grows." What they oppose is a pig-
headed pro-growth economic policy that facilitates bigger cities,
a larger population, the consumption of more fuel and electricity,
and the production of more stuff. Inevitably, the centralized "giantism" denounced by Schumacher favors big businesses, pollutive fuel
sources, and a zealous worship of technology, while turning a blind
eye to the lamentable corollaries of "more."

Mishan asserts that economic growth was essentially invented
in the nineteenth century but only took off in the 1940s. By the
1960s, the National Economic Development Council in England,
the hugely influential and international Organisation for Economic Co-operation and Development, and numerous countries
in the Western world had made economic growth an official part
of governmental policy.[19] That is, the decision to stimulate growth
was *made by economists and politicians*—it did not occur as some
kind of "natural" historical accident. Given the role that a pro-
growth policy played in exacerbating the world's ecological crisis
and given the propensity for growth to impede a high quality of life
for future generations, the ecological economists determined that
this policy needed to be abandoned. Thus Mishan stated: "Certainly there can be no purely economic justification for a policy of
growth *per se*."[20] For Daly, favoring growth was hypocritical and

irresponsible: "A policy of maximizing GNP is practically equivalent to a policy of maximizing depletion and pollution."[21]

It is important to keep in mind that this debate took place against the backdrop of high unemployment and slow growth in the 1970s, in addition to new research about mounting ecological crises and the scarcity of resources—all of which emboldened the ecological economists to stress the drawbacks of growth addiction. Daly, in particular, used the "failed growth economy" as the springboard for his proposals for a steady-state economy. This debate also took place during the startling oil crises that opened eyes to the fragility of the global economy. In October of 1973, the Arab member-states of OPEC punished the United States for supporting Israel in the Yom Kippur War by imposing an oil embargo that drove up prices in the West until the embargo was lifted in 1974. Lovins cites the 1973 crisis as the inspiration for his work on the "soft energy path."[22] A second major disruption to oil exports occurred in 1979, too, when the turmoil surrounding the Iranian Revolution slowed the oil sector and drove up prices.

But in a broader sense, the "crisis" of the 1970s was the growing realization that the age of cheap and abundant crude oil was quickly coming to an end. M. King Hubbert's warnings about "peak oil" production in the United States—which he estimated would occur between 1965 and 1970 in the lower forty-eight states—proved rather accurate.[23] Peak world production of oil would come in the early twenty-first century, but it hardly bode well for those global "exponential-growth culture[s]" (Hubbert's term) that relied so heavily on nonrenewable fossil fuels. The good times of oil-fueled-growth capitalism apparently couldn't last forever.

Mishan's 1977 *The Economic Growth Debate: An Assessment* reflects on and documents the economic and political uncertainty of the late 1970s. In that book, Mishan responds to the three main pro-growth arguments that remained a part of economic "conventional wisdom" (a term popularized by John Kenneth Galbraith).[24] The three arguments were as follows: (i) that a scarcity of resources will cause prices to rise and that people will respond by using less of a resource and searching for new ones; (ii) that markets bring "optimal time paths" to the use of nonrenewable resources, which allows, again, for society to locate new resources; and (iii) that past concerns about declining resources have been unwarranted. Mishan refutes each argument in turn: Assertion #1 does not solve the problem of scarcity; assertion #2 is based on abstractions and blind faith in the wisdom of markets; and assertion #3 does not mean that we can or should repeat the same rates of resource depletion that we've caused over the past 200 years.[25] Daly, for his part, adds to these arguments by stating that rampant growth, which ultimately backfires in the long run, has been the problem in industrial society, and thus it cannot simultaneously be the solution.[26]

The final argument against growth posited by ecological economists in the 1970s centers on social factors. Mishan contends that economic growth has not generated an equal distribution of wealth and income in British society. "Indeed," he argues,

> We might go so far as to suggest that economic growth *per se* should be jettisoned as an independent goal of policy. For if we are concerned primarily with social welfare, those forms of economic growth that meet our welfare criteria will in any case be approved and adopted, the remainder being rejected.[27]

Daly concurred that boosting overall wealth in a society did not promote equality or cure social ills—in fact, it created new ones. "The moral facts are asserting themselves," he wrote in 1977, "in the form of increasing existential scarcity: anomie, injustice, stress, alienation, apathy, and crime."[28] Fred Hirsch, in his 1976 *Social Limits to Growth*, argues that social limits might even outweigh biophysical ones: "The concern with the limits to growth that has been voiced by and through the Club of Rome is strikingly misplaced. It focuses on distant and uncertain physical limits and overlooks the immediate if less apocalyptic presence of social limits to growth."[29]

Together, the ecological economists argue that most of the wealth that has been created in the twentieth century has gone up the social ladder, creating huge gaps in inequality in industrialized societies. Moreover, they contend that growth that favors the rich effectively breaches the stated principles of justice and equality that, in theory, guide Western societies. These critiques about social class and social consequences represent some of the first attempts on the part of economists to bring together social, environmental, and economic considerations—an important development for the sustainability movement that coalesced in the 1980s.

If not growth, then what? It is in the responses to this question that we see the greatest differences between the ecological economists. None of them put forth a fully operational proposal for abandoning a growth economy and establishing a sustainable economic system, although Daly perhaps came the closest by sketching out a macroeconomic framework for a new economy. (More detailed proposals would come later from sustainability economists.) But the ecological economists did point the way to new economic goals and systems that would slow environmental degradation and promote social justice. Dennis Clark Pirages argues in favor of a slow-growth economy that would help create a "sustainable society"—his edited book is called *The Sustainable Society: Implications for Limited Growth* (1977)—by "conserving energy and other natural resources" and developing the solar power industry.[30] Mishan wanted forms of non-growth "development" that advanced "social welfare," from education and healthcare to the arts and leisure time. He also thought that a slow growth rate did not have to jeopardize a high standard of living—if one counts "frequent tea-breaks and other manifestations of disguised leisure" as "goods," then Britain has benefited from slower growth.[31] Schumacher invented something called "Buddhist economics" that would value people, labor, and the environment over wealth and material consumption.[32]

Other economists were more vocal, however, in desiring a no-growth economy, even if that meant governmentally imposed regulations that *prevented* growth (in throughput and population, most notably). If the growth economy was a conscious construction, they argued, then it could be consciously deconstructed. In his 1976 *The Sustainable Society: Ethics and Economic Growth*, Robert L. Stivers joins Paul Ehrlich and Barry Commoner in advocating for "Zero Population Growth," but he extends the argument to include "Zero Economic Growth."[33] Daly is also in favor of a no-growth economy, at least when it comes to the industrialized world; he finds it hypocritical to reject at least some economic growth in poorer countries.[34] What Daly proposes is a "steady-state economy" that takes its inspiration from John Stuart Mill's 1848 *Principles of Political Economy*, the iconoclastic work of political economy discussed in chapter 2 that advanced the idea of a "stationary state of capital."[35] Daly defines a steady-state economy as a "low energy economy" with "constant stocks of people and artifacts, maintained at some desired, sufficient levels by low rates of maintenance 'throughput.'"[36] Democratically elected officials would be in charge of implementing maximum resource quotas and population control initiatives:

> We need (1) an institution for stabilizing population (transferable birth licenses); (2) an institution for stabilizing the stock of physical artifacts and keeping throughput below ecological limits (depletion quotas auctioned by the government); and

(3) a distributist institution limiting the degree of inequality in the distribution of constant stocks among the constant population (maximum and minimum limits to personal income and a maximum limit to personal wealth).[37]

Daly is adamant that the steady-state economy would allow for moral and cultural "development" but not economic growth and that a steady-state economy is different from a "recession" or a "failed growth economy."[38] Even though some of Daly's more extreme suggestions have been set aside as impractical, such as his proposals for aggressively limiting population growth, his notion of a steady-state economy continues to inform the economics of sustainability in the present day as does his notion of uneconomic growth. At the very least, Mishan, Daly, Schumacher, the Club of Rome, and the other first-wave ecological economists succeeded in making growth, limits, and societal overshoot legitimate topics of debate.

THE NATURAL ENVIRONMENT IGNORED

A second ingredient in the critique of neoclassical economics was that it purposefully ignored the environment or that, at best, it viewed it as an inert entity that was essentially external to the concerns of wealth creation. The ecological economists accused mainstream economists of treating nature like a huge dumping ground for industrial pollution. They also faulted the neoclassical economists for assuming, naïvely, that the law of supply and demand would drive up prices on resources that became scarce, thus slowing consumption rates and allowing nonrenewable resources to last almost indefinitely. To counter that argument, the ecological economists pointed to the disappearance or degradation of many natural stocks as examples of how free markets did relatively little to prevent overconsumption. While economists blathered on, they suggested, the Earth's finite resources became ever scarcer.

Moreover, they denounced the way in which neoclassical economics "externalized" the natural world from economic equations. That is, conventional economic theory assumed that the natural world was "out there" and thus not inherently part of economic systems. Nature entered the economic realm only when an item gained "exchange value" as a commodity. Trees and soil and waterways have no inherent value in neoclassical economics, despite the invaluable ecosystem services that they provide. In an ordinary cost–benefit analysis, and assuming an absence of environmental regulations, a business would not be obliged to count toxic sludge dumped into a lake or pollution spewing from a smoke stack as a "cost"—in fact, in many ways, it's seen as a benefit. Once a pollutant enters the "commons," it is no longer the concern of the business, even if that sludge or air pollution causes human health issues or environmental catastrophe. The ecological economists found this situation untenable.

In fusing ecology and economics into a single discipline, they sought to devise new ways of integrating environmental factors into economic theory. For them, the thinking of the neoclassical economists was redolent of a bygone worldview that assumed that nature could not be seriously harmed by innocuous human technologies. Two pioneers in bridging this gap were Howard T. Odum, the renowned American ecologist whose concept of "energetics"—the study of energy flows—had both ecological and economic components, and C. S. Holling, a Canadian ecologist whose work on the "resilience" of ecological systems had an impact on both science and economics.[39]

What were known as "external diseconomies" in the 1960s and "externalities" in the 1970s were all those costs, as Mishan put it, that economists opted conveniently to neglect. Instead of calling pollution what it was, it became euphemized as "spillovers" or the "spillover effect."[40] In other words, the bad stuff was nudged out of the model even though it led to "social conflict," health problems, and eco-disasters.[41] Mishan argues that economic theory in his day lacked methods for understanding the impact of these various costs: "Some sorts of external diseconomies, manifestly important ones at that, do not lend themselves easily to measurement—no small defect in a society so prone as ours is to equate relevance with quantification."[42] He gives some examples of difficult-to-calculate social and environmental costs:

> Some of the simple nuisances, on the other hand, such as excessive engine noise and emission of noxious fumes, may be tackled most economically by enacting compulsory noise-muffling measures and compulsory installation of anti-fume devices, as in several states of the US. However, more general social afflictions such as industrial noise, dirt, stench, ugliness, urban sprawl, and other features that jar the nerves and impair the health of many are difficult both to measure and to impute to any single source—which is, of course, no reason for treating them with resignation.[43]

One of the problems with externalities, exposed by Kenneth Boulding, is that they imply, fallaciously, that one can have a boundless, consequence-free economic system. Boulding terms this reckless concept the "cowboy" economy, to which he compares the "spaceman" economy that operates responsibly within a single "econosphere."[44] Drawing on the second law of thermodynamics (entropy), he argues that the new economy must be thought of as a kind of "closed" system in which the atmosphere and the oceans are not seen as "external." That is, as Hardin and other ecological thinkers had noted, there is no "away" to which we can throw things. "In the spaceman economy," Boulding continues, "what we are primarily concerned with is stock maintenance, and any technological change which results in the maintenance of a given total stock with a lessened throughput (that is, less production and consumption) is clearly a gain."[45] The famed economist Nicholas Georgescu-Roegen made similar arguments about the thermodynamic foundation of economic processes in *The Entropy Law and the Economic Process* (1971), thus helping to bridge economics and the science of energy transformation.[46] Odum, too, argued that "the economist must learn how energy sources work. He calls such sources *externals* and often is unaware that their flows control the economy and cannot be ignored."[47]

One means of rectifying the problem of external diseconomies was to "internalize" formerly externalized costs into economic equations. Certainly, some economists defended this idea in the 1970s. But there were problems with internalizing the external, too. As Mishan noted, there was a paucity of methods for quantifying many externalities within mainstream economics. For Daly, the issue was not so much methodological as it was ecological: "Internalizing externalities into relative prices deals only with relative scarcity, not at all with absolute scarcity. Orthodox economics has treated all scarcity as relative, so naturally it considers internalization of externalities to be the whole answer. But of course it is not."[48] Not only did internalizing externalities fail to deal with absolute declines in nonrenewable resources, which was Daly's focus in this passage, but it also did nothing to prevent ecosystem destruction, pollution, and all those social costs of growth discussed above. (At best, it just made consumption more expensive.) In short, the ecological economists sought new measures and metrics that could capture the real cost of economic activity instead of simply setting aside social and ecological realities. The "true cost economics" that has come to flourish in the present day owes quite a lot to these early critics of externalities.

USELESS METRICS AND MEASUREMENTS

A third avenue of criticism ran straight toward what ecological economists saw as the worthless indicators of economic well-being that neoclassical economics unthinkingly employed. Virtually all of the ecological economists mercilessly deride the use of the gross domestic product (GDP)/gross national product (GNP) as a useful measure. In textbook terms, GDP measures the total market value of a country's products and services. GNP also measures the total value of goods and services but modifies the GDP by adding into the equation total capital gains from overseas investments and income from citizens living abroad (but it does not count the income of domestic nonresidents). In the words of a recent sustainability economist, GDP/GNP measures only the "busy-ness" of an economy and crudely assumes that more activity is necessarily better.[49] The first-wave ecological economists essentially make the same argument, sometimes to hilarious effect. As Stivers puts it, "GNP is roughly analogous to an overweight man. He may not want or need all the pounds he puts on."[50]

The problem with this economic obesity was that it gave the false impression that a corpulent economy was a healthy and resilient one. Even Simon Kuznets, who invented the GDP metric in 1937, warned that it was subject to "illusion and resulting abuse," mainly because it did not account for "the personal distribution of income" or "a variety of costs that must be recognized."[51] Mishan, in the 1960s, was one of the first to criticize the GDP/GNP for that very reason: It was a faulty metric of economic health that ignored the costs of economic activity. One of its many shortcomings is that it purports to be neutral when in fact it is a covertly normative index that is biased in favor of certain values and desires—mainly the desire for more consumable stuff.[52] Neoclassical economists basically always see growth in GDP/GNP as a positive development.[53] But this growth often comes at the expense of the natural environment and even rewards pollution, waste, inefficiency, and the depletion of nonrenewable resources.[54] This ubiquitous metric, ecological economists noted, is so ethically impoverished that it perversely counts environmental disasters as a good thing, since environmental cleanups are profitable for certain businesses. The now-classic example of this perverse logic that a disaster is good for society (Frédéric Bastiat in 1848 called this idea the "Broken Window Fallacy") is the Exxon Valdez oil spill off the coast of Alaska in 1989—a disaster of unparalleled scope that ended up boosting the US GDP by at least $2 billion.[55]

Moreover, the GDP/GNP fails to take into account uneconomic growth—the idea that some growth is bad and harms society in the long run—since these metrics ignore the distant future and qualitative analyses.[56] Long-term issues pertaining to standards of living are set aside in this quest for present-day "busy-ness." "The problem with GNP," Daly observed, "is that it counts consumption of geological capital as current income."[57] What Schumacher, Daly, and Mishan sought was an economic metric that valued not only the conservation of resources but also social welfare and equality. Mishan wrote in 1977, somewhat idealistically perhaps, that "even the most conservative economist now agrees that changes in GNP are no longer acceptable as an index of changes in social welfare or in social product."[58]

Instead of the GDP/GNP, which basically measured economic growth, the ecological economists wanted metrics that valued things like equality and resource conservation, took qualitative factors into consideration, and penalized uneconomic growth. At the very least, argued Boulding, the GNP should distinguish between renewable and nonrenewable resources so that economists can track problematic growth.[59] Stivers mentions some of the early efforts on the part of economists to create new indices for social well-being, including the Measure of Economic Welfare (created in 1972) and an early "Development Index," both of which jettison mere quantitative analysis and develop qualitative measures of a society's economic health.[60] Thus, as we've seen over the past few decades as alternatives to the GDP/GNP continue to develop, there are many scenarios in which GDP/GNP growth—clearly seen as a good thing—can be considered *quite a bad thing* in a rival metric. Daly also offered some rudimentary alternatives to the GNP/GDP, but along with his peers, called for greater research in alternative metrics.

TECHNOLOGY WORSHIP

A fourth criticism of neoclassical economics is that, by promoting "more" at all costs, it assumes that man-made technologies are inherently benign or at least "worth" the collateral damages. This criticism is basically identical to the environmentalist critique of the narrative of "progress" that the twentieth century inherited from the Industrial Revolution and the Enlightenment. In the words of Schumacher, our "greatest successes" have, in fact, done the most to threaten the natural environment and the human economy that relies upon it.[61]

The most obvious "successes" were in the energy sector. Both Schumacher and Lovins argued that nuclear power was not worth the risks. Not only did it not solve the problem of meeting growing electricity demands, but also it created numerous hazards for the environment and political stability. Bear in mind that these warnings were penned *before* the catastrophic nuclear accident at Chernobyl. Schumacher asks how it is possible to ensure that nuclear waste can be stored safely for 25,000 years—the time that it takes to reach its radioactive half-life.[62] Even more troubling, the construction of nuclear reactors threatened global peace efforts since the same materials that went in to making nuclear power— uranium and plutonium—were also the core elements of nuclear weapons. There is a reason that Lovins subtitled his book "Toward a Durable Peace"—because he recognized that the "hard" technologies (i.e., environmentally impactful ones), especially nuclear fission and fusion but also the fossil fuels, brought with them warmaking capabilities.[63] Both he and Schumacher published their respective books in the mid-1970s, at the height of the nuclear arms race, when the United States and the Soviet Union had each spent trillions on building nuclear missiles and warheads that numbered in the tens of thousands. In a way, Lovins and Schumacher were part of a tradition of ethical scholars that went back to Albert Einstein, who regretted, later in life, his involvement in the creation and promotion of atomic weapons.

The other "hard" energy sources condemned by ecological economists were fossil fuels: crude oil, coal, and natural gas. Relatively little was known about fossil-fueled anthropogenic climate change in the 1970s, so none of these economists based their critiques on the heat-trapping capabilities of carbon dioxide emissions. That argument came later. Due to the research of Barry Commoner and others, however, they did understand the potential for carbon emissions to create smog, soot, and "noxious fumes" that made it hard on the lungs of those who lived in mega-cities or near industrial zones. Of equal or greater concern was Hubbert's warning about peak oil and the inevitable price hikes and shortages that would come with it. Odum, too, made it clear in the 1970s that economic growth was tied directly to finite fossil fuels: "Energy controls growth."[64] Odum, Lovins, and Schumacher wanted to transition away from technologies that relied on fossil fuels before fuel shortages forced industrial society to grind to a halt—and also before society made the gigantic step backward from scarce crude oil to the more plentiful and more pollutive sedimentary coal. A third problem was that fossil fuel tended to rely on large-scale, inefficient, centrally controlled technologies—for the refining, distribution, and consumption of these fuels.[65] Thus oil and coal tended to go hand in hand with technologies that undermined equality and made it harder for individuals to live self-sufficiently.

Instead of "hard" energy and its attendant technologies, Lovins provides a detailed blueprint toward the "soft energy path." Drawing inspiration from Schumacher, he argues that soft energy and technologies differ drastically from the more recognizable hard ones. They are based on renewable energy flows; efficient; flexible, low tech, and environmentally friendly; diverse and dispersed (rather than uniform and centralized); and produced so that scale and energy quality meet end-use needs.[66] The main soft energies touted in this period for electricity production were photovoltaic solar power and wind turbines, both of which were still relatively new technologies in the 1970s. The US government also explored algae-based biofuels via the Aquatic Species Program founded in 1978 under the Carter administration.

There is a distinct flavor in the work of Lovins and other energy specialists that the soft energy path could be the only means for humans to survive what seemed like an inevitable, apocalyptic breakdown of industrial society. Indeed, survivalism was a popular cause in the 1970s and drove interest in weaning the developed world from fossil fuels. Many people in this period wanted to live in homes or communities that produced their own food and energy, and they viewed self-sufficiency as a form of freedom. The modern sustainability movement clearly owes quite a lot to these pioneers who advocated for communal autonomy and researched "alternative energy."

Ecological economists also argued that the neoclassical attitude toward "the new" and modern technology often lacked an ethical component. It's not a coincidence that the subtitle to Stivers's book is "Ethics and Economic Growth." The ecological economists had deep-seated concerns about the "creative destruction" (Schumpeter's famous term) that defined modern capitalism. There seemed to be an assumption in the twentieth century that new gadgets and techniques, as long as they increased productivity or short-term human comfort, were somehow inherently better than the old ways—even if the new ways meant more pollution and resource consumption. The ecological economists were not simpleminded technophobes. Rather, they wanted a society that prioritized moral and ecological considerations over short-term economic ones and evaluated the myriad consequences of material innovation. Important questions needed to be asked: Does technology X allow humanity to live within its biophysical limits? Is it overly pollutive or destructive? Are the benefits worth the costs, defined broadly? (In a sense, the latter had been the Luddite's question 150 years earlier.)

Those were the sorts of value-laden questions that neoclassical economists often set aside in the hunt for "the new." After all, greed, shortsightedness, and apathy toward morality had undoubtedly led to the creation of the "throw-away society" (a term invented by *Life* magazine in 1955). Both Daly and Schumacher wanted a rigorously ethical economic system that fostered human development over waste and inequality. Daly argues:

> We need . . . to shift the emphasis toward ecological adaptation, that is, to accept natural limits to the size and dominion of the human household, to concentrate on moral growth

and qualitative improvement rather than on the quantitative imperialist expansion of man's dominion. The human adaptation needed is primarily a change of heart, followed by a shift to an economy that does not depend so much on continuous growth.[67]

Early ecological economists put forth a comprehensive critique of business-as-usual economics. They wanted to make the "dismal science" of economics more ethical and more centered on ecological knowledge. They foresaw a future in which growth was set aside as a governmental policy and population stabilized at a manageable figure. They also envisioned a society that fostered education about the natural world and accepted strict environmental regulations and taxes, all in an effort to construct an equitable human *oikos* that operated safely within its biophysical limits. They wanted to renew and reorient the study of economics, which had apparently lost sight of the ultimate purpose of an economy. Even though these economists met with limited success between the 1960s and early 1990s, they have greatly informed the sustainability movement of recent decades. Indeed, as we will see, the core features of today's economics of sustainability have grown straight out of the works of the early ecological economics.

Chapter 5

From Concept to Movement

"From growth to sustainability."

—Lester R. Brown, 1981

A self-defined sustainability movement crystallized between the late 1970s and the 1990s. No longer was sustainability merely a concept or set of ideas. There was now a set of organizations—the Worldwatch Institute, the Rocky Mountain Institute, the United Nations (UN), and so on—that promoted something called "sustainability" and a growing number of individuals who sought to "live sustainably." Scholars began to describe in vivid detail what a sustainable society might look like and discussed in no uncertain terms the unsustainability of modern industrial society. In 1975, a conference was held near Houston, Texas, on "how a modern society might be organized to provide a good life for its citizens without requiring ever-increasing population growth, energy resource use, and physical output."[1] A stream of books between 1976 and

1981 drew on cutting-edge science and ecological economics to sketch out the "qualitative components of a sustainable society."[2] In the 1980s, sustainability became the centerpiece of international agreements; a strategy objective for at least some nongovernmental organizations (NGOs), businesses, and governments and a philosophy of balance and durability with a wide range of applications. It found its greatest champion in the United Nations though, which recast sustainability as "sustainable development" and integrated its principles into international accords.

Sustainability had thus become part of a political agenda and a clearly articulated ecological philosophy, and a plethora of frameworks, systems, and models were developed as a means of studying, measuring, and advancing its central tenets. This is the period, for instance, in which the three Es appeared as the basic model for sustainability. By the 1990s, sustainists had begun implementing principles of sustainability into economic analyses, planning commissions (on all governmental levels), the energy sector, education, agriculture, housing, transportation, business operations, and many other domains. The media picked up on the term, too, and sustainability became, by the end of the century, a buzzword meant to signify anything associated with green values.

This chapter offers a brief overview of the formation, triumphs, and challenges of the sustainability movement at the end of the twentieth century. As we have seen, though, the sustainability movement did not appear spontaneously around 1980. The academics, diplomats, and activists who furthered the "quest for a sustainable society" drew on two decades of research on overlapping subjects: pollution, environmental degradation, and climate change; the limits to population, consumption, and throughput growth; the need to stabilize biological systems; the potential to tap renewable energy sources and increase energy efficiency and conservation; and the prospects of building sustainable cities and agricultural systems. Of course, the champions of sustainability stood on the shoulders of giants from an earlier age. There would not have been a sustainability movement without John Evelyn, Jean-Baptiste Colbert, Hans Carl von Carlowitz, Jean-Jacques Rousseau, Thomas Malthus, John Stuart Mill, Charles Darwin, George Perkins Marsh, John Muir, Rachel Carson, Jane Jacobs, and countless other men and women who developed ecological thinking.

Yet most scholars agree that the watershed moment for sustainability occurred in 1972 with the publication of the Club of Rome's paradigm-shifting book, *The Limits to Growth*. In the words of one scholar, "Into this complacent world [modern industrialism] came the staggering hypothesis that the world system would suffer catastrophic collapse before the year 2100 unless drastic movement was begun immediately toward reaching a state of global equilibrium."[3] Another scholar notes, "The word *sustainable* appear[ed] for the first time in its modern, broader meaning in" the Club of Rome's report, which sought a world system "capable of supporting human life" in the long term.[4] The club depicted sustainability as a philosophy of social stability and the antithesis of suicidal growth. In a very real sense, the sustainability movement that took shape institutionally in the 1980s was a constructive response to the challenges mounted by the Club of Rome and other participants in the "growth debate" of the 1970s. The movement was and remains an attempt to construct a sustainable society and avoid the societal collapse envisaged by the most well-informed systems theorists of the twentieth century.

It is quite striking to read the books and conference proceedings on sustainability from the late 1970s and early 1980s. There is a genuine sense of hopeful optimism that the environmental disasters and economic hardships of the past decade would bring about a sudden transformation toward social equilibrium. Here's one early sustainist in 1981: "Something very important happened in the 1970s. . . . What happened was that we . . . changed our minds about growth."[5]

But it became quickly apparent that only a tiny minority of people had changed their minds about growth—at least among the leaders of the industrialized world. It turned out that the 1980s would belong to Ronald Reagan and Margaret Thatcher. Neoliberal economic policies in many parts of the developed world brought about the deregulation of financial markets and the privatization of many public services. The Soviet Union teetered and collapsed, providing a massive public relations coup for capitalism and the "free world." Materialism enjoyed an almost spiritual revival. Money was worshipped. Emissions and population and consumption and throughput all increased dramatically. Just as it had been since World War II, economic growth was championed even when it was detrimental to social well-being and the environment.

The 1980s was thus marked by a polarized view of industrial capitalism. On the one side stood the Reaganites and Thatcherites, who backed the West's standard of living and military might and who warmed their hands in the smoldering embers of the Soviet bloc. The idea of limits fell by the wayside and fossil fuels were embraced liked never before. Perhaps the most emblematic event of the period came in 1981, when Reagan removed the solar panels from the roof of the White House that President Carter had so gladly and symbolically installed. On the other side stood a range of social and environmental groups who envisioned a different path for society, one that was not rooted in materialism, growth, fossil fuels, and mounting social inequalities. Some very big things seemed to hang in the balance in the mid-1980s—modernity and the "progress" of humanity, neoclassical economics, the very fate of industrial society—but, argued the sustainists, those with the power to make change had instead revitalized the very political and economic policies that had fostered unsustainability. As a result, the defenders of sustainability had hardened their tone by the end of the decade; the naïve optimism had given way to regret and defensiveness. Thus

Gro Harlem Brundtland, in her foreword to the hugely influential *Our Common Future* (1987), lamented the "marked retreat" from social and environmental concerns that characterized the 1980s.[6] The point here is that sustainability did not magically win over the hearts and minds of business and government leaders in the 1980s. In fact, progress in building a sustainable society was rather slow, and the successes in this period, although notable, were more significant for their symbolic potential than their immediate impact.

Perhaps the greatest success of the sustainability movement in the 1980s, however, was the fact that high-ranking diplomats within the UN latched on to the idea and made it a central aspect of international accords. Many of these diplomats came from countries that had been badly affected by ecological problems and that happened to lack the entrenched interests of the fossil fuel industry. Beginning in the 1970s, the UN stepped up its involvement in coordinating international environmental efforts and assisting member states in implementing environmentally sound practices and policies.

The turning point, in this regard, came with the aforementioned UN Conference on the Human Environment held in Stockholm in 1972. Although the concept of sustainability was not explicitly discussed at Stockholm, two events of lasting consequence occurred during the conference. The first was the creation of the hugely influential United Nations Environment Program (UNEP), which became the permanent environmental branch of the UN. The second was the promulgation of a "Stockholm Declaration," which included key principles related to the "human environment." The declaration is rather convoluted and at times bizarrely worded, but it clearly incorporates ecological thinking. Proclamation 6 from the Declaration states: "A point has been reached in history when we must shape our actions throughout the world with a more prudent care for their environmental consequences. Through ignorance or indifference we can do massive and irreversible harm to the earthly environment on which our life and well-being depend."[7] The declaration set a precedent for later UN declarations of sustainability that would have a tangible impact on global environmental policy and development schemes.

An important moment for sustainability came in 1980, when UNEP commissioned the International Union for the Conservation of Nature (IUCN), an organization of some 700 scientists from over 100 countries, to write a comprehensive report that would "help advance the achievement of sustainable development through the conservation of living resources."[8] The resulting *World Conservation Strategy: Living Resource Conservation for Sustainable Development* is notable for a number of reasons. First, it was apparently the first international document to use the term "sustainable development," which resonated with the UN's mandate to promote the well-being of impoverished countries but which also elicited criticism from many sustainists who objected to this seemingly contradictory term. Second, the report is important because it reflects a shift in environmental consciousness from strict conservationism—the protection of resources and the environment—to a more constructive philosophy of social transformation and a more dynamic appreciation of the interplay between the environment, the economy, and human well-being. That is, one sees quite clearly in this document the emergence of the sustainability discourse and a move away from an older and narrower conception of environmentalism. Consider here the IUCN's systemic approach to development:

Develop is defined here as: the modification of the biosphere and the application of the human, financial, living and non-living resources to satisfy human needs and improve the quality of human life. For development to be sustainable it must take account of social and ecological factors, as well as economic ones, of the living and non-living resource base; and of the long term as well as the short term advantages and disadvantages of alternative actions.[9]

The shift to ecological systems thinking that we see here has remained part of the UN approach to development ever since 1980.

Third, the report is important because of its frank message to policymakers, conservationists, and development practitioners. The three stated goals of the report are to maintain essential ecological processes and life-support systems, to preserve the planet's genetic diversity, and to ensure the sustainable use of species and ecosystems. Along the way, it addresses climate change (in vague terms), depletion of the ozone layer, desertification (the spread of deserts), the health of the world's oceans, and many other scientific subjects. However, it is clear that, despite the IUCN's awareness of the interrelationship between environmental, economic, and social problems, the main emphasis here is on ecosystem science, and the report says relatively little about social justice, poverty, inequality, women, the indigenous, faulty economic and financial systems, and other subjects addressed in later UN documents on sustainable development.

Significantly, the emergence of sustainability within the milieu of the UN signaled the increasingly international character of the movement. This book has focused on the Western world, but from the 1980s onward, the quest for sustainability was a thoroughly globalized endeavor. The participants in UN conferences on sustainability, as well as the authors of the many declarations and treaties that surfaced from these meetings, came from all over the developed and developing world. For instance, the Peruvian Secretary-General of the UN, Javier Pérez de Cuéllar, established the World Commission on Environment and Development (WCED) in 1983 to create a global framework for sustainable development. The chairman of the WCED was Gro Harlem Brundtland, the former (and future) prime minister of Norway; the vice chairman was a Sudanese diplomat named Mansour Khalid; and the principal author of the document produced by the WCED—*Our Common Future*—was a Canadian environmentalist and the Director of Environment for the Organisation for Economic Co-operation and Development named Jim MacNeill. The UN became the locus of this internationalized movement because, as *Our Common Future* acknowledged, the entire globe faced the interlocking problems of environmental degradation, resource scarcity, overpopulation, and social inequality.

Indeed, *Our Common Future* represents the next major step for the sustainability movement. This report is still frequently cited as the Ur-text for sustainability, even though, as we've seen in this book, both the concept and the movement antedate 1987 by quite some time. When the UN created the WCED, they tasked it with creating a viable strategy for "achiev[ing] sustainable development by the year 2000" and made the charismatic Brundtland the chairperson of the committee.[10] Indicative of the shifting environmental consciousness of the day, the initiators of the WCED understood clearly that environmental problems were related directly to social, economic, and development issues and that "the environment" could no longer be approached in a vacuum. As Brundtland notes in her foreword, there were those who wanted the WCED to limit itself solely to "environmental issues," but the committee organizers rejected this idea on the grounds that the environment did not exist as a separate sphere from human actions.[11] "There has been a growing realization in national governments and multilateral institutions that it is impossible to separate economic development issues from environment issues." In fact, *Our Common Future* probably deserves much of the credit for establishing the three Es as the basic model for a sustainable society. There is an acute awareness within the text that environmental problems are indelibly linked to poverty, overpopulation, and the unequal consumption of resources. That is, there is a true balance of concern for all three of the Es of sustainability.

The report also deserves recognition for its pithy and widely reproduced definition of sustainable development (which certainly gets more air time than the richer one offered in 1980 by the *World Conservation Strategy*): "Humanity has the ability to make development sustainable to ensure that it meets the needs of the present without compromising the ability of future generations to meet their own needs."[12] To be sure, countless organizations and scholars have adopted *Our Common Future*'s intergenerational definition of sustainable development, and the report's balanced approach to sustainability has set the tone for international development policy since the late 1980s. Table 5.1 summarizes the major international conferences, commissions, declarations, strategies, and treaties between 1969 and 2012 that contributed to the sustainability movement.

As the sustainability movement grew and became more diversified in the 1980s, however, it became plainly obvious that it was suffering from an identity crisis. The UN conferences of the decade made it clear that policymakers often disagreed radically on approaches to sustainability. The structure of the UN and the interests of its member states meant that sustainability began to take on new characteristics, some of which were at odds with the concept that had taken shape in the 1960s and 1970s.

The first important alteration stemmed from the top–down nature of UN procedures. Rather than embracing Schumacher's conception of decentralized decision making and local autonomy, the UN instituted a hierarchical and centralized approach to sustainability, based on the lowest common denominator of environmental

Table 5.1 INTERNATIONAL COOPERATION AND THE GROWTH
OF SUSTAINABILITY, 1969–2012: NOTABLE CONFERENCES,
COMMISSIONS, DECLARATIONS, STRATEGIES, AND TREATIES

1969, San Francisco, United Nations Educational, Scientific and Cultural Organization (UNESCO) conference called "Man and His Environment: A View Towards Survival"

The conference welcomed 500 attendees and focused on environmental issues, survivalism, food security, and the threats of nuclear power. The conference did not address sustainability explicitly but set a precedent for United Nations (UN) involvement in international environmental discussions.

1972, Stockholm, UN Conference on the Human Environment and the "Stockholm Declaration"

This conference is considered a watershed moment in the history of international cooperation on environmental issues. It led to the creation of the UN Environment Program (UNEP) and formulated a "Declaration" with 7 proclamations and 26 principles. The declaration deals with the safeguarding of the natural environment and social issues such as colonialism and oppression. Also, it was around this time that the UN began discussing the relationship between human-caused air pollution and global climate change.

1979, Convention on Long-Range Transboundary Air Pollution

This international agreement, which was signed by numerous countries, has been expanded over time by a series of protocols, and represented an early attempt to reduce and prevent air pollution across international borders. It set the stage for UN involvement in climate change policy.

continued

Table 5.1 (continued)

1979, Vienna, UN Conference on Science and Technology for Development

This conference, which also resulted in a program of action, was not particularly concerned with the environment, but it did bring to prominence issues surrounding international development, which would play a central role in the UN conferences of the 1980s and 1990s.

1980, UNEP-backed *World Conservation Strategy: Living Resource Conservation for Sustainable Development*

This document was created by the International Union for Conservation of Nature, which was made up of hundreds of scientists and researchers from around the world. It is apparently the first international document to use the term "sustainable development." The report was commissioned by UNEP, but additional support and input came from the World Wide Fund for Nature, the Agricultural Organization of the UN, and UNESCO. This well-written report deals with environmental and development issues and urges world societies to "come to terms with the reality of resource limitation and the carrying capacity of ecosystems."

1980, Brandt Commission releases *North-South: A Program for Survival*

The former Chancellor of Germany, Willy Brandt, led an international commission that studied development inequalities between the global north and the global south. Its report, released in 1980, focuses on poverty and environmental issues. It discusses sustainability and draws a distinction between "development" and "growth."

1982, UN General Assembly, "World Charter for Nature"

The United States, threatened by the pro-environment language, was the only country to vote against this charter, which was adopted by the UN General Assembly. It acknowledges that "mankind is a part of nature and life depends on the uninterrupted functioning of natural systems." It also gestures toward more explicit statements of sustainable living in passages such as this one: "Man can alter nature and exhaust natural resources by his action or its consequences and, therefore, must fully recognize the urgency of maintaining the stability and quality of nature and of conserving natural resources."

1985, Vienna Conference and the "Vienna Convention for the Protection of the Ozone Layer"

The Vienna Conference was an international meeting meant to address the expanding hole in the stratospheric ozone layer, the cause of which was the release of man-made refrigerants, such as CFCs and halons, into the atmosphere. These compounds depleted the ozone layer, allowing more UV light to enter Earth's atmosphere, which increased rates of skin cancer in humans and caused a range of environmental problems. The "Vienna Convention" was a non-binding multilateral agreement that laid out a framework for dealing with CFCs.

1987, Montreal Protocol on Substances that Deplete the Ozone Layer

The protocol which was brokered by UNEP, expands upon the Vienna Convention. It is a legally binding treaty that has been signed by all UN member states and which deals with the phasing out of chlorofluorocarbons (CFCs) and other ozone-depleting compounds. It has been amended several times since 1987. The protocol

continued

Table 5.1 (continued)

has been called the greatest success in international environmental cooperation, and because of the phaseout of CFCs, the ozone layer has already begun to reconstitute itself.

1983–1987, the World Commission on Environment and Development (WCED) and *Our Common Future* (published in 1987)

The UN established the semi-independent WCED in 1983. It was chaired by Gro Harlem Brundtland, and due to her strong leadership on the commission, it has since been known as the "Brundtland Commission." The purpose of the commission was to create a framework for global sustainable development. The document produced by the commission is called *Our Common Future*, although it is also known as the "Brundtland Report." It puts forth a definition of sustainable development that is still cited today: "It meets the needs of the present without compromising the ability of future generations to meet their own needs." It also states very explicitly the interrelationship between environment, economics, and social issues. The influence of this report is almost incalculable.

1988, Intergovernmental Panel on Climate Change (IPCC)

The UNEP and the World Meteorological Organization created the IPCC in 1988. The IPCC was and remains an association of scientists dedicated to analyzing and summarizing current research on anthropogenic climate change. The IPCC has released "assessment reports" for diplomats and the public in 1990, 1992 (a supplement), 1996, 2001, 2007, and 2014 (the Fifth Assessment Report). The reports have become more forcefully worded over time in assessing the role that humans have played in changing the planet's climate, primarily through the emission of greenhouse gases (GHGs).

1988, Toronto, Group of Seven (G7) Summit

Although little came from it, the Group of Seven industrialized countries discussed anthropogenic climate change and carbon dioxide emissions, which set a precedent for G7 and later G8 discussions on the environment.

1992, Rio de Janeiro, UN Conference on Environment and Development (known as the "Rio Earth Summit") and several policy documents: "Forest Principles," "Agenda 21," "Rio Declaration on Environment and Development," "Convention on Biological Diversity," and the "Framework Convention on Climate Change."

The Rio Earth Summit was a huge and well-publicized conference attended by UN member states and thousands of NGOs. The "Rio Declaration" contains 27 principles that were meant to guide policy toward the environment and development. Principle 3 basically restates the main definition of sustainable development in *Our Common Future*. "Agenda 21" offers a detailed framework for implementing sustainable development. The "Framework Convention on Climate Change" (UNFCCC) and the "Convention on Biological Diversity" were the only two legally binding treaties to emerge from the conference, but many states declined to sign on. The UNFCCC was later superseded by the Kyoto Protocol (1997), which established legally binding emissions reduction obligations for developed countries.

1993, UN Commission on Sustainable Development

The General Assembly created this commission in 1993 to oversee the implementation of Agenda 21 and other measures from the Rio Earth Summit. It lasted until 2012, at which point a new Division of Sustainable Development was created.

continued

Table 5.1 (continued)

1997, Kyoto Protocol to the UN Framework Convention on Climate Change

The Kyoto Protocol is an environmental treaty that was adopted in 1997 (and became enforceable law in 2005) and which stipulates that developed countries (that have signed and ratified the treaty) reduce GHG emissions in an effort to curb anthropogenic climate change. It covers two periods of targeted emissions reductions: 2008–2012 and 2013–2020. Currently, 37 countries are bound to the treaty, but Canada and the United States are not directly involved. The treaty established new emissions trading systems and credits for countries that finance emissions reduction programs in the developing world. The treaty has helped many countries, especially in Europe, reduce GHG emissions.

2000, UN-backed "Earth Charter"

An international commission run by Maurice Strong, who had over-seen the Rio Summit, and former head of the Soviet Union Mikhail Gorbachev facilitated the drafting of an "Earth Charter." The charter raises "ecological sustainability" to the level of a global ethic. It also ties sustainability to peace efforts, social equality, and respect for all life. It was the international community's most confident statement to date on the need for a "sustainable global society."

2002, Johannesburg, World Summit on Sustainable Development (known as "Earth Summit 2002") and "Johannesburg Declaration on Sustainable Development"

This seldom-discussed meeting was meant as a 10-year reunion for the Rio Earth Summit of 1992. The "Johannesburg Declaration" is a rather tepid restatement of earlier declarations on sustainable development. The United States, led by President George W. Bush, boycotted the summit, weakening the environmental credibility of both the United States and the UN.

2009, Copenhagen, UN Climate Change Conference

A disastrous climate change conference beset by strong disagreements about how to reduce GHG emissions. The Copenhagen conference is a strong indicator of the inability of developing and developed countries to reach a consensus about how to move forward on climate change. No agreement or meaningful plan of action was adopted by the conference attendees.

2012, Rio de Janeiro, UN Conference on Sustainable Development (known as "Rio + 20") and *The Future We Want*

The Rio + 20 conference was the 20-year reunion of the Rio Earth Summit. Much of the conference was marked by lamentations about failed efforts to implement the kind of sustainable global order envisioned back in 1992. *The Future We Want* strikes a defeatist tone and speaks of "renewing political commitment" to the cause of sustainability. The conference also led to the creation of a new Division of Sustainable Development within the UN Department of Economic and Social Affairs.

consensus among member states—and with decidedly mixed results. The second change was that the UN transformed sustainability into "sustainable development." It might seem like hair-splitting semantics, but there were some important differences between the concepts of sustainability and sustainable development. The systems theorists, environmentalists, and ecological economists of the 1970s, after all, had not focused very much on "development," concentrating instead on rethinking growth-based economics and establishing a society that lived within ecological limits. By contrast, the phrase "sustainable development" seemed to imply something different.

What was wrong with sustainable development? The problem, from the standpoint of many sustainists, was that the term was, at best, dangerously ambiguous, and at worst, masked ulterior motives. When the word "sustainable" entered the English language in an economics dictionary in 1965, it was attached to the word growth—"sustainable growth"—and was meant to denote the gradual and continual expansion of the GDP, the perennial dream of neoclassical economists. The problem that many scholars have pointed out is that "development" seems like little more than a euphemistic synonym for "growth," and if that is the case, then "sustainable development" is not a concept that favors a steady-state economy or ecological stability but is rather a covert and greenwashed vehicle for business-as-usual economic policy.[13] In the words of John A. Robinson, the issue is that "development is seen as synonymous with growth, and therefore that sustainable development means ameliorating, but not challenging, continued economic growth."[14]

Robinson argues that sustainable development suffers from three conceptual pathologies: It is "vague, attracts hypocrites and fosters delusions." His contention is that sustainability and sustainable development, although often conflated and used synonymously, represent, in fact, two different strands of environmental thought in the Western world: sustainability derives from John Muir's notion of "preservationism," whereas sustainable development traces its roots to Gifford Pinchot's pro-business conception of "conservationism."[15] Likewise, the physicist Albert A. Bartlett has argued that "sustainable development" and "sustainable growth" were, in fact, used interchangeably by the time of the Clinton–Gore administration, despite engendering an obvious oxymoron: if something is being "sustained," it is not actively "growing" or "developing."[16] It became apparent to many sustainists in the 1980s that sustainability had become a Hydra-like concept and that the central contention within the movement turned on the question of growth.

One can see this inconsistent approach to growth and economic values in the UN documents from the 1980s. It's quite apparent that multiple authors with conflicting views of sustainability took part in these policy-writing efforts since patently contradictory views of growth are frequently expressed within the same document and sometimes even within the same run-on sentence. UN diplomats and the researchers who worked with them could not decide whether a sustainable society would come about through deregulated financial systems and growth-based economics or through regulated economic systems and the cessation of pro-growth policies. Thus, the *World Conservation Strategy* presses for a "new international economic order" and the need to "regulate international trade" to ensure that resources in developing countries aren't exploited and that consumption globally remains at "sustainable levels." However, a few pages later the report touts neoliberalism and suggests that "trade be liberalized, including the removal of all trade barriers to goods from developing countries."[17]

Our Common Future is even more blatantly contradictory. At first, it defends the idea of ecological limits and slams economic growth for creating environmental ruin, industrial pollution, and the overconsumption of resources in the industrialized world: "Environmental concern arose from damage caused by the rapid economic growth following the Second World War." Yet the report then performs an about-face and urges the need for a "new era of economic growth." At one point, the authors even use the dreaded term "sustainable economic growth" that had caused so much consternation among anti-growth sustainists.[18]

So, is economic growth the problem or the solution? As Herman Daly noted in the 1970s, it cannot simultaneously be both. The UN's ambivalence toward growth has led to accusations, since the 1980s, that sustainable development is merely neoclassical economics in disguise, and economic growth, indeed, remains a significant point of contention within the world of sustainability.[19]

Perhaps the discord within the sustainability movement accounts for the rather mixed results that the UN achieved between 1980 and 2000. Let's begin with some of the successes. The first achievement, as noted, was that the UN put sustainability on the map by giving the movement an institutional home. Sustainability became in the 1980s a legitimate development framework and an objective to which global societies could aspire.

Second, UNEP could point to some signal achievements in international environmental policy. It not only commissioned the *World Conservation Strategy*, but it also brokered perhaps the world's greatest triumph in international environmental cooperation—the Montreal Protocol on Substances that Deplete the Ozone Layer (1987). This legally binding treaty outlawed chlorofluorocarbons (CFCs) and other ozone-depleting compounds, and all UN member states eventually signed on. As a result, the ozone layer has already begun to heal itself, which means fewer cancer-causing UV rays are able to reach human bodies on the surface of the planet.[20] UNEP also cocreated the Intergovernmental Panel on Climate Change (IPCC), which has been responsible for summarizing current research on anthropogenic climate change since 1988. The periodic assessment reports of the IPCC have guided climate change policy around the globe. Global societies have been aware of the ways in which humans can affect weather and climate, but it wasn't until the late 1960s and 1970s that links began to be drawn between carbon dioxide emissions and global warming.[21] Ever since the 1980s, the sustainability movement has been closely associated with growing concerns about climate change and its effects on humans and ecosystems.

Third, the UN organized well-publicized summits that produced influential treaties and environmental policies. The most important development conference was the Rio Earth Summit of

1992, in which member states and hundreds of NGOs met in Rio de Janeiro, Brazil, and discussed environmental and development issues. Out of the summit came several binding and nonbinding agreements. The "Rio Declaration" cast sustainable development for the first time as a human "right," which meant that living sustainably could now be seen as a natural entitlement rather than a hopeful objective. The wording created a legal basis from which to argue that unsustainability was not only against the law but even a human rights violation.[22] Another policy document, Agenda 21, was (and remains) a nonbinding but very detailed framework for countries and local communities to develop sustainably. Thousands of towns and regions have implemented Agenda 21 into sustainability action plans, resulting in decreased emissions, energy conservation, more sustainable urban planning, and decreased resource consumption. Agenda 21 has been particularly influential in Finland, the United Kingdom, and Sweden.[23] The Earth Summit also gave birth to the legally binding "Framework Convention on Climate Change" (UNFCCC) and the "Convention on Biological Diversity."

The UNFCCC was later superseded by the Kyoto Protocol of 1997, which established legally binding emissions reduction obligations for the developed world. Although adopted in 1997, the Kyoto Protocol did not become law until 2005. Currently, 37 states are bound to the treaty, although Canada and the United States are not. Both of these industrial powerhouses fear the short-term effects that emissions reduction efforts would have on their respective economies. The protocol stipulates that developed countries must reduce greenhouse gas (GHG) emissions during two periods—2008 to 2012 and 2013 to 2020—in an effort to slow or reverse anthropogenic climate change.[24] During the first period, many countries involved in the treaty, especially in Europe, were able to meet their emissions reduction targets. The treaty is also noteworthy because it established new "emissions trading systems" and credits for industrialized countries that finance emissions reductions programs in the developing world.[25]

On balance, however, UN involvement in sustainability has suffered from many shortcomings and failures. The main issue is that many of the treaties, frameworks, and agreements mentioned above have been ineffectual and are often seen as a load of hot air. The Rio Summit was marked by bitter disagreements between the global North and the global South and between NGOs and governments. The documents produced in Rio were greatly watered down and reflected the disagreements of member states. The following UN summits in Johannesburg (2002) and Rio (2012) basically acknowledged the massive gap between where the UN would like the world to be and where it actually is. Global inequalities remain widespread, the world's population continues to grow, and the consumption of many scarce resources, including fossil fuels, remains a stumbling block to the creation of a sustainable society and a green economy. The UN has, in fact, done little to transition the world away from growth-based economics and neoliberalism. Despite some progress on battling climate change, GHGs continue to build up in the atmosphere, and carbon dioxide levels surpassed 400 parts per million in 2013, bringing the world ever closer to runaway climate change.

One of the central complaints about the UN is that many of the agreements, such as Agenda 21, are voluntary and nonbinding. Even the binding legal agreements, such as the Kyoto Protocol, allow for member states to bow out at any time, which is what Canada did in 2011. Even more troubling is the fact that UN-brokered climate accords have focused narrowly on developed countries and are basically powerless to reduce emissions in China, India, Mexico, Brazil, and other "developing countries." Moreover, the politically left-of-center and top–down nature of the UN has led to a backlash from some conservative polities. The United States is famously ambivalent about the UN and has not ratified the Kyoto Protocol, and the State of Alabama even went so far as to legally forbid the implementation of Agenda 21, mainly because of its pro-environment and pro-labor policies.

Nevertheless, the 1980s and 1990s witnessed growing efforts, both inside and outside the UN, to turn sustainability into action. Whereas the ecological economists of the 1970s had been rather short on constructive ideas, sustainists of the late twentieth century developed and implemented numerous plans, initiatives, and detailed strategies to bring a sustainable society to fruition.

Comprehensive schemes appeared throughout this period outlining what would constitute a sustainable society and how one might go about implementing and rating it. From the 1970s onward there were attempts to make sustainability less vague and more operational, all in the hopes of answering that ever-pressing question: How can humans live sustainably upon the Earth? A series of "Woodlands Conferences," held near Houston, Texas (1975, 1977, and 1979), invited dozens of scholars to address this very question. The essays presented at the conferences offered a range of practical ideas, including the development of alternative fuels (biomass, solar, geothermal, wind, etc.) and the shift to "voluntary simplicity."[26]

In 1981, Lester R. Brown, the founder of the Worldwatch Institute, crafted what might have been the first detailed blueprint for constructing a sustainable society. His *Building a Sustainable Society* analyzes the core problems of industrial society and maps out a viable "path" to sustainability as well as some of the "institutional challenges" that stand in the way. The core of his strategy focuses on steadying the population, preserving natural resources, stabilizing biological systems, developing renewable energy, constructing sustainable transportation and agricultural systems, creating green jobs and an educational system that prepares people for them, rethinking urban planning, and establishing "greater local self-reliance."[27] The institutional challenges that he discusses include overcoming vested interests—especially in the fossil fuel and industrial sector—and rallying businesses, universities, public interest groups, and the media to the cause of sustainability.[28] In the same vein, in 1989, David Pearce, Anil Markandya, and Edward B. Barbier published a pathbreaking work on developing a "green economy."[29]

The still-vague concept of sustainable development also took off in this period, moving beyond the UN and toward international financial institutions and universities. In the 1990s, universities throughout the industrialized world—especially in Canada, the United Kingdom, and the United States—began offering under-graduate and graduate courses and even degrees in sustainable de-velopment.[30] It became an academic field and an area of expertise. Moreover, international financial institutions, such as the World Bank, ordinarily a bastion of neoliberalism and environmental apathy, picked up on sustainable development as a way to bridge environmental and development issues. "It is now recognized that a healthy environment is essential to sustainable development and a healthy economy," the institution noted in 1992.[31]

But the World Bank wanted to get beyond the ambiguities of the concept and thus began researching practical means of measur-ing whether a development scheme was sustainable. A World Bank report from 1992 called *Sustainable Development Concepts* uses the "conventional neoclassical theory of economics" to analyze, somewhat surprisingly, how "free market forces may not achieve sustainability, and how policy intervention may help or hinder sustainability."[32] Similar to the contradictions found in the UN, however, the World Bank in the 1990s used business-as-usual eco-nomics to criticize the effects of business-as-usual economics. It differentiated between, yet blended together, "sustainability" and "sustainable economic growth"; a mixture of "government intervention" (regulation) and "market mechanisms" are touted as the best means to "improve sustainability."[33]

By the end of the twentieth century, however, it was clear to many scholars that sustainable development, while expanding as a discourse within international development, politics, and higher education, remained divided between two opposing camps:

> One advocates continuing economic growth, made much more environmentally sensitive, in order to raise living standards globally and break the links between poverty and environmental degradation. The other calls for radical changes in economic organization, producing much lower rates of growth as we know it, or even zero or negative growth.[34]

Although conventional "market-based approaches" might have dominated the theory of sustainable development in the 1990s, progress was made in the opposing camp, too. New ideas emerged about "fair trade" as opposed to "free trade" and emphasis was placed on local governance (as opposed to the top–down approach of the UN and the World Bank), sustainable agriculture, environmental and human health, women, the indigenous, and education issues.[35] In a sense, there was a growing conflict between the idea of "sustainable development" and the notion of *developing toward sustainability*—that is, moral, cultural, and economic transformation that aimed for ecological stability.

The last decades of the twentieth century also witnessed a growing interest in sustainable, organic agriculture. Although basically all food from 10,000 BC to around 1900 was grown "organically," with natural fertilizers and pest-control techniques, the twentieth century gave birth to industrial forms of agriculture: the abandonment of ancient forms of crop rotation and the advent of chemically fueled monocropping that relied on synthetic fertilizers, pesticides, and herbicides, many of which were made from petroleum and which polluted groundwaters, depleted and toxified soils, and killed off birds and bees.[36] The organic food movement took shape in the 1970s, in part as a reaction to Rachel Carson's condemnation of pesticides, but it really took off in the 1990s. Since 1990, sales of organic foods in the United States have increased by an astounding 20% annually. By 2005, consumer sales had reached $13.8 billion and continue to grow today.[37] Organic farming, which has been legislated and regulated in many parts of the world, is an "ecological production management system that promotes and enhances biodiversity, biological cycles, and soil biological activity" and is "based on minimal use of off-farm inputs." In the United States, this means that organic foods are "grown without synthetic pesticides, growth hormones, antibiotics, modern genetic engineering techniques (including genetically modified crops), chemical fertilizers, or sewage sludge."[38] Although organic farming and the use of non-genetically modified crops has long been criticized by international development organizations on the grounds that it yields less food for the world's poor, recent developments have shown that organic farming techniques can actually outproduce conventional farming methods and without the same risks to water and soils.[39]

One approach to organic farming that has grown to prominence around the world is permaculture. In the 1970s, the Australian ecologists Bill Mollison and David Holmgren developed permaculture as a response to the devastating effects of industrial agriculture in Tasmania, where deforestation and chemically fueled monocropping had depleted soils, polluted waterways, and reduced biodiversity. Permanent agriculture, or permaculture, began as a form of agricultural design but later expanded into an all-encompassing philosophy of sustainable living, which stresses effective farming techniques and community self-reliance.[40] The 12 design principles of permaculture, which have varied slightly over time, bring together many of the ecological ideas that are central to the sustainability movement:

1. Observe and interact;
2. Catch and store energy;
3. Obtain a yield;
4. Apply self-regulation and accept feedback;
5. Use and value renewable resources and services;
6. Produce no waste;
7. Design from patterns to details;
8. Integrate rather than segregate;
9. Use small and slow solutions;
10. Use and value diversity;
11. Use edges and value the marginal; and
12. Creatively use and respond to change.

One can glimpse here how these principles might be applied to more than just farming or gardening, although agriculture remains the primary application of permaculture practices. Permaculture has become quite popular in many parts of the world since the 1980s and remains for many people their primary connection to sustainability. Their ideas became a blueprint for getting society away from a style of food production based on animosity, chemicals, and exhaustion and toward one based on mutualism, renewable resources, and continued fertility. Mollison and Holmgren helped inspire, for instance, the widespread use of mulching, composting, rainwater collection, and the "harmonious integration of aquaculture, horticulture, and small-scale animal operations."[41]

Another development was the fair trade movement, which emerged in the latter decades of the twentieth century as a means of creating more sustainable forms of global trade and consumption. Although the roots of fair trade stretch back into the mid-twentieth century, by the 1970s and 1980s, the movement had become a viable alternative to neoliberal globalization policies and "free trade," which created unfettered markets that tended to exploit goods producers and degrade ecosystems in the developing world. Fair trade, by contrast, did not value low prices above all else and instead sought to create greater equality between consumers in the developed world and goods producers in the developing world. Fair trade advocates believe that trade should be based on higher environmental standards—trade is not supported if it depends upon environmentally destructive practices—and the notion that producers and exporters of goods—especially small-scale, impoverished, and historically exploited artisanal craftsworkers—should be able to charge higher-than-normal prices for their goods, so that they can subsist at a respectable standard of living.

Alternative trading organizations, such as the Oxfam Fair-Trade Company, began to appear in the United Kingdom and North America and became important advocates for the fair trade industry. The movement originally centered on handicrafts, but textiles, coffee, chocolate, fruits, and other agricultural commodities came to dominate fair trade late in the century.[42] Fair trade still constitutes a small percentage of global trade, but it has produced many positive social and environmental impacts. The creation of fair trade labels has helped solidify the market for these goods, which continues to expand, along with organic foods, at an astonishing rate, and total annual retail sales of fair trade products now count in the billions.[43]

Although the environmental movement initiated the recycling movement, as we saw in chapter 3, recycling has since become part of sustainability's goal of eliminating waste and conserving energy. Of course, as with organic farming, recycling is nothing new, and every pre-industrial society relied on some form of resource recycling. The generation that lived through the Depression and World War II recycled extensively, too. What was new between the 1970s and 1990s was the attempt to intensify the recycling of industrially produced materials (glass, paper, aluminum, steel, electronics, etc.) and to make recycling a serious and permanent way of reducing energy consumption. The extracting of metals from ore, after all, is very expensive and energy intensive. It is difficult to calculate how much total material and energy has been conserved because of recycling efforts since the 1970s. What's easier to determine is the growth in recycling rates. Rates in America jumped from 9.6% of municipal garbage in 1980 to 32% in the early twenty-first century. Some countries in Europe now recycle over 60% of municipal waste.[44] In 1995, Americans recycled 47.6 billion soft drink containers. Indeed, the recycling industry grew steadily from the 1980s onward, as cities around the world began picking up recyclables at the curb and as thousands of new recycling centers opened and helped divert the waste stream. In the 1990s, "reduce, reuse, recycle" became a mantra for every schoolkid, and many cities began to, in effect, require recycling in an effort to reduce waste. Government at all levels began to mandate the purchasing of recycled goods (especially paper). Although recycling has been called the greatest success of the environmental movement, and it remains an important aspect of the quest for sustainability, the ongoing critique is that it has not been able to reduce overall consumption rates in the industrialized world. "As the volume of waste has increased, so have recycling efforts."[45]

The growth of renewable and sustainable energy from the 1970s to the 1990s is another indicator of the growing influence of the sustainability movement. As with recycling and organic food, renewable energy has a long history. The wind, trees, animals, and humans that powered pre-industrial societies were all "renewable." Dutch windmills that pumped water or milled grain harnessed a renewable form of energy. So, too, did sustainably harvested forests, which provided an endless supply of burnable wood (biomass). In fact, basically all energy was renewable before the Industrial Revolution brought fossil fuels into the equation. As we have seen in this book, one of the major causes of unsustainability in the industrialized world has been the use of nonrenewable coal, oil, and gas. Thus the rediscovery of renewables in the late twentieth century is something of a "return to origins," even though new technologies modernized the old forms of energy. Many sustainists in this period, including Amory Lovins and Lester R. Brown, devised schemes for transitioning industrial society away from "hard" and unsustainable fuels to "soft" and renewable ones.[46]

In the last two decades of the twentieth century, wind power, solar power, biomass, biofuels, and geothermal energy (the "soft") began to make a dent in the consumption of industrial energies (the "hard"): gasoline, diesel, and electricity produced from fossil fuels and unsustainable nuclear power. They joined hydropower, which was a renewable (although not necessarily sustainable) form of electricity generation that had existed since the late nineteenth century. Many of these renewables were developed in response to the energy crisis of the 1970s, when the reliance on foreign oil became devastatingly obvious. The United States began installing wind turbines that generated electricity that fed the electrical grid. Three large wind farms appeared in Southern California in the early 1980s.[47] Biomass power plants that burned plant matter to generate electricity appeared in many countries.

Solar energy, in the sense of harnessing the heat and light of the sun, had been used in a range of different ways in pre-industrial societies (e.g., in greenhouses). But solar *power* converted sunlight into electricity using photovoltaics and, more recently, via concentrated solar power. Similarly, solar thermal energy uses solar energy to heat water. Early solar technologies began to develop as early as the 1860s but mostly stalled until the 1970s. Photovoltaics sales ebbed and flowed between the 1970s and 1990s in an inverse relationship with oil prices, although the amount of solar-generated gigawatts of electricity has grown steadily since 2000. In 1977, the path-breaking Saskatchewan Conservation House, which used a solar heating system, became a model for the "net zero" homes of the twenty-first century.

Geothermal energy, which uses heat from the Earth to generate electricity and heating, became common in many countries, including the United States and Iceland, both of which have the geothermal activity to support this energy source. Between 1970 and 2000, Iceland shifted almost entirely from oil-based heating to geothermal heating. The fuel crisis also spurred interest in biofuels in Brazil, Canada, the United States, Europe, and elsewhere. The feedstock for biofuels can come from sugarcane (ethanol), corn (ethanol), canola (biodiesel), soy beans (biodiesel), and many other sources, although serious questions remain about the sustainability and scalability of these fuels. From 1978 to 1996, the United States studied the science and scalability of algae-based biodiesel, which many today believe offers a more sustainable form of biofuel.[48]

Despite these advances, the use of renewables still paled in comparison to hard energy consumption by the end of the twentieth century. In some places, hydro provided a significant source of electricity, but fossil fuels, even more so than nuclear power, remained the primary motive force in the world. Wind and solar power played bigger roles in global energy consumption only after 2000. It became increasingly clear around the turn of the century, as oil production began to peak (but consumption continued to grow), that no single renewable would emerge as the universal remedy to oil addiction. The prevailing logic since 2000 is that global societies will need to draw on available local, decentralized, and renewable energy sources and use them in conjunction to definitively displace hard energy.

The first serious efforts to transport the principles of sustainability into the world of business and commerce occurred in the 1980s and 1990s. Even in the late 1970s, participants in the Woodlands Conferences had urged business to get on board with the new ecological thinking.[49] In 1981, Lester Brown argued that one of the central stumbling blocks to creating a sustainable society would be overcoming the vested interests of corporations that profited from unsustainable industrial practices.[50] Yet increasingly in this period, attempts were made to harmonize the interests of corporations and the cause of the environment. John Elkington, Karl-Henrik Robèrt, Paul Hawken, Amory and Hunter Lovins, and others sought new ways of making capitalist enterprise more ecological and sustainable, while still prioritizing the need for profitability. These writers remain important in twenty-first-century efforts to establish a "green economy," and all of them have helped stimulate the growth of environmental consulting firms.

Karl-Henrik Robèrt was a pioneer in this regard. Born in Sweden, Robèrt rose to prominence as a cancer researcher in the 1980s. He became increasingly concerned with the buildup of toxins within bodily tissue and sought to use his knowledge of cell biology to advocate for human health and sustainability. He drew on his scientific credibility to broker a consensus among the Swedish scientific community about what conditions would need to be met to ensure the existence of a sustainable society. The Swedish scientists agreed on four systems conditions for sustainability, which took on somewhat different forms over the years. But the conditions in the 1990s usually looked like this:

1. "In the sustainable society, nature is not subject to systematically increasing concentrations of substances extracted from the Earth's crust."
2. "In the sustainable society, nature is not subject to systematically increasing concentrations of substances produced by society."
3. "In the sustainable society, nature is not subject to systematically increasing degradation by physical means."
4. "In the sustainable society, human needs are met worldwide."[51]

Robèrt called his system The Natural Step and set up a consulting agency that could help businesses and other organizations realign their values with those of sustainability. Robèrt did not take issue with capitalism or money-making, and he didn't seem particularly concerned with the problem of growth. Rather, he saw The Natural Step as a means of rethinking the ultimate purpose of business. Using a method called "backcasting," his company began helping businesses figure out how to implement the use of sustainable materials and practices. Major clients have included IKEA, McDonald's of Sweden, Fabriks, and Scandic Hotels. The Natural Step now has offices in Sweden, the United States, the United Kingdom, Canada, Australia, New Zealand, Japan, Israel, and South Africa and remains an important bridge between the business world and sustainability.

Paul Hawken was an early collaborator with Robèrt. He set up a branch of The Natural Step in the United States after working for many years as an environmental consultant. In 1993, he published what would become a very influential book on the need to overcome the notion that "commerce and sustainability were antithetical by design."⁵² *The Ecology of Commerce: A Declaration of Sustainability* offers a blueprint for managing a profitable *and* environmentally responsible business. "The question is, can we create profitable, expandable companies that do not destroy, directly or indirectly, the world around them?"⁵³ Unlike many of the ecological economists of this period, Hawken did not take direct aim at economic growth or a deregulated capitalist economy. Aside from advocating for taxes on consumption and pollution, his book stresses the need for adherence to "market principles." What he wanted was for businesses to promote sustainability by profiting from it—to green themselves and, in the process, turn a healthy profit.

That is not to say that businesses should profit from just anything. Hawken laments the fact that the business world has only two words for profit—gross and net—and doesn't distinguish between responsibly generated profits and profits that derive from practices that harm people and the environment.⁵⁴ For Hawken, "the ultimate purpose of business is not, or should not be, simply to make money. Nor is it merely a system of making and selling things. The promise of business is to increase the general well-being of humankind through service, a creative invention and ethical philosophy."⁵⁵ The book offers a list of six practices that a sustainable business would follow:

1. "Replace nationally and internationally produced items with products created locally and regionally."
2. "Take responsibility for the effects they have on the natural world."
3. "Do not require exotic sources of capital in order to develop and grow."
4. "Engage in production processes that are human, worthy, dignified, and intrinsically satisfying."

5. "Create objects of durability and long-term utility whose ultimate use or disposition will not be harmful to future generations."

6. "Change consumers to customers through education."[56]

Hawken's ecological approach to capitalism has been continuously discussed since the 1990s. His book is seen as a palatable and non-threatening entryway into sustainability by corporations that might otherwise see sustainability as dangerous and anticapitalist. This approach certainly works *with* business, not against it, in an effort to reorient the economy away from pollution and destructive growth.

Hawken later teamed up with Amory Lovins, the environmental scientist and energy expert, and Hunter Lovins, the writer and sustainability advocate, to write the best-selling book *Natural Capitalism: Creating the Next Industrial Revolution* (1999). Amory Lovins had been well known since the 1970s, when he wrote *Soft Energy Paths,* and the Lovinses had run a successful research and consulting firm called the Rocky Mountain Institute since 1982. The book that they wrote together expands on many of the ideas in *The Ecology of Commerce* but also develops Schumacher's idea of "natural capital," which, for these authors, refers to the inherently valuable "sum total of the ecological systems that support life."[57] The book argues that a new, more sustainable form of capitalism could emerge if ecological limits are respected and natural capital is seen as inherently valuable, even before it is turned into an exchangeable product.

As with the works discussed above, *Natural Capitalism* is fundamentally favorable to "market-based systems of production and distribution" but seeks a more ecologically sensitive form of capitalism. This book has become famous for its assumption that society could become sustainable not through wholesale reconsideration of human values but simply through technological fixes and workable alterations to the current economy. Hawken, Lovins, and Lovins devised four strategies for creating natural capitalism:

1. Radical resource productivity: Getting more out of resources, especially energy resources.
2. Biomimicry: Modeling the economy on natural processes, eliminating the concept of waste, and reusing materials in "continuous closed loops."
3. Service-and-flow economy: Shifting the economy from "goods and purchases" to one in which customers lease products and services from producers, who have an incentive to produce durable goods and offer superior service.
4. Investing in natural capitalism: Restoring and supporting ecosystems.[58]

The book's shining example of a natural capitalist enterprise was Interface Inc., which was (and remains) one of the world's largest manufacturers of commercial carpet. Owned by the eco-minded Ray C. Anderson, Interface committed to sustainability in the early 1990s, drastically reducing (and even eliminating) its waste, increasing its resource efficiency, committing to a service-and-flow model, and employing biomimicry to create mix-and-match flooring designs that created new efficiencies within the company. It also made a lot of money.

Natural Capitalism addresses many other subjects, too. Along the way, it predicts the rise of hybrid vehicles, makes a strong pitch for resource efficiency, and urges businesses to "tunnel through the cost barrier," which is the idea that focusing on long-term energy efficiency will save money down the road. The attention received by this book was a boon for the Rocky Mountain Institute, which has done environmental consulting work for dozens of Fortune 500 companies.

Perhaps the most recognizable advent in business sustainability during this period was the idea of the "triple bottom line" (TBL). Introduced in 1997 by the corporate responsibility expert John Elkington in his groundbreaking *Cannibals with Forks: The Triple Bottom Line of 21st Century Business*, TBL is the idea that a business should aim for and be able to measure success in three complementary arenas: economic (profits), social (social well-being and justice), and environmental (sustainability and environmental quality).[59] Elkington thus echoed Hawken's assertion that the purpose of a business is not simply to make money; businesses must promote social justice and incorporate the principles of sustainability if they hope to remain viable and profitable in the long term. The first bottom line of the TBL is economic, which takes into consideration the status of human, physical, and financial capital. The second bottom line is environmental. The idea here is that profits should be achieved by respecting "natural capital" and through the use of renewable energy. The third bottom line is social. A company should value quality and well-being and try to close the gap between the rich and the poor. It should promote health and education. It should not profit from human suffering or undertake practices that exacerbate social problems.[60]

Since the 1990s, TBL has become a common philosophy in the world of corporate responsibility, and many companies, mostly small- and medium-sized ones, have implemented it in their normal accounting practices.[61] Although no major, publicly traded company is currently using TBL, medium-sized companies such as Patagonia have employed it to good effect. Patagonia, a recognizable outdoor apparel and accessory company, posted annual sales of $540 million in April of 2012. Patagonia used TBL's slogan of "people, planet, profits" to shift, for instance, toward the use of organic cotton and recycled soda bottles (which can be turned into fleece). The company not only supported organic cotton farmers by committing to buying their products but also made renewed efforts to increase employee satisfaction and well-being.[62]

Finally, the field of ecological economics continued to develop in the 1980s and 1990s, as the views of Schumacher, Daly, and others slowly percolated their way into the world of economics. In 1989, the field received its own journal in the shape of *Ecological Economics*. A handful of ecological economists in this period, including Daly himself, found work in banks and international financial institutions, such as the World Bank.[63] But the most important development in the field was undoubtedly the emergence of a concept that has continued to shape the economics of sustainability: Ecological Footprint Analysis (EFA).

The idea of an "ecological footprint" was developed by the ecological economist William Rees and his Swiss graduate student at the University of British Columbia, Mathis Wackernagel. Together, Rees and Wackernagel produced a book and a series of articles in the 1990s that provided a solid method for measuring the sustainability of a given political unit. EFA measures human demands on the Earth's ecosystems. It shows how modern societies tend to use ecological services faster than Earth can renew them. More precisely, it measures how much land and sea area is needed to support the lifestyle and consumption rates of a given human unit—say, a city or a country. "The ecological footprint of a specified population or economy can be defined as the area of ecologically productive land (and water) in various classes—cropland, pasture, forests, etc.—that would be required on a continuous basis" to provide the material, energy, and pollution-processing needs of a given polity.[64] It doesn't say much about social justice or other social factors pertaining to sustainability, but it is quite effective at showing human stresses on the natural environment.

The ecological footprint is important for several reasons. First, it defines the concept of a "city" or a "country" along ecological lines. A city is not merely an area of land surrounded by a border but rather a consumptive unit that relies on ecological goods and services from around the globe. As Rees put it, "All urban regions appropriate the carrying capacity of distant 'elsewheres,' creating dependencies that may not be ecologically or geopolitically stable or secure."[65]

Second, Rees and Wackernagel formulated a precise methodology to generate a specific, measurable, and useful numerical value for the land (and sea) demands—that is, the ecological footprint—of a given political unit. For example, with the city of Vancouver, "to support only the food and fossil fuel demands of their present consumer lifestyle, [Vancouver's] 1.7 million people require, conservatively, 8.3 million hectares of land in continuous production. The valley, however, is only about 400,000 hectares. Thus, our regional population 'imports' at least 20 times as much land for these functions as it actually occupies."[66] The ecological footprint thus served as a quantitative means of showing that if every city or country consumed as much as, say, Vancouver, the world would be in a state of ecological overshoot. If the whole *world* consumed as much as Vancouver does, then we would need three planet Earths to survive.[67]

Third, the ecological footprint became a powerful way to criticize neoclassical, growth-based economics and the faults of industrial globalization. Rees and Wackernagel took the insights of the first-wave ecological economists and turned them into a viable method that could be used as a basis for ensuring that societies remain within their ecological limits. Rees and Wackernagel have made "footprints" a ubiquitous word in contemporary society. In addition to the ecological footprint, there is also now the "carbon footprint," which measures the GHG emissions of an individual, organization, or political unit. But what was so significant about

EFA was that it generated numbers that showed how unsustainable our world currently is. For these two researchers, sustainability isn't some vague and unmeasurable concept. Rather, it is possible to show concretely that many cities and countries are living beyond their available ecological means.

EFA proved that the world has been in a state of ecological overshoot since the end of the twentieth century. Now, what governments and the public do with that information is up to them. Certainly Rees and Wackernagel want global societies to reduce their ecological footprints and function more sustainably. To this end, Wackernagel serves as the president of the Global Footprint Network, a think tank that uses EFA to help cities and countries reduce pollution and consumption rates. The organization has done consulting work with such polities as Calgary, San Francisco, London, Milan, the European Union, Japan, Costa Rica, and the United Arab Emirates.[68] The ecological footprint remains widely used and discussed and has helped shift sustainability from a vague idea to a precise methodological designation and a viable academic field.

By the turn of the twenty-first century, sustainability was firmly entrenched within academia, the UN, and numerous governments, NGOs, and businesses. In 2000, Switzerland became the first country in the world to incorporate "sustainable development" into its constitution.[69] In a short 25-year period, sustainability had gone from an amorphous and reactive idea to a constructive movement supported by powerful institutions. Again, this is not to say that the world suddenly became sustainable but rather that new attention was given to the faults of growth and industrialism. The Bhopal industrial disaster (1984) and the Chernobyl nuclear disaster (1986) served as stunning reminders that industrial pesticides and hard energies were dangerous and unsustainable.[70] Moreover, new data on mankind's role in climate change sparked concerns about the ultimate cost of industrial society. Was overshoot and collapse in our future?

The sustainability movement certainly grew in this period and gained new methods and applications, from agricultural systems to business and economics. But it was also beset by contradictions and divisions. The disagreements over growth between the backers, respectively, of sustainability and sustainable development remained an endemic problem for the movement. To this day disagreements persist between those who seek sustainable development and those who want global society to develop toward sustainability. The paradoxes and mixed results of the UN in its promotion of sustainability were emblematic of the fate of the movement as the 1990s drew to a close.

Chapter 6

Sustainability Today: 2000–Present

"Successful gardens do not keep expanding."

—David Holmgren

"Absent any adaptation, systems that follow a single exponential growth curve inevitably collapse."

—Andrew Zolli and Ann Marie Healy, citing Geoffrey West

We might not live in a sustainable age, but we're living in the age of sustainability. The movement has gained a level of prominence in recent years that is difficult to dispute. The scholarly fields associated with sustainability have expanded dramatically; new tools and methods have appeared that help define, measure, and assess sustainability; and a broad range of organizations and communities have embraced the principles of sustainable living. Sustainability, in fact, has gone from marginal ecological idea to mainstream movement in a surprisingly short amount of time. We now see sustainability publicized at the supermarket, on university campuses, at the aquarium, in corporate headquarters, in government ministries, and in countless other places. A growing number of universities, nongovernmental organizations (NGOs), and corporations in the Western world possess an "office of sustainability"—replete with sustainability plans and guidebooks—but none have an "office of green radicalism" or an "office of the status quo." In a sense, this environmental discourse has won out over rival conceptions of humanity's relationship to the natural world.

This chapter is an attempt to sketch out the different ways in which sustainability has gained a foothold in contemporary society. It is not meant to suggest that our world is sustainable. On the contrary, many barriers and entrenched interests have kept our world rather unsustainable, and Mathis Wackernagel has even argued that, since the 1990s, we have exceeded the Earth's capacity to sustain us; we are now living in a state of global overshoot.[1] The goal here, rather, is to show the ways in which our society has constructively responded to our ecological crisis—to demonstrate the growth and elaboration of the sustainability movement and describe some of the successes it has achieved in counteracting our bad habits.

As the philosophy of sustainability has developed, so too has it expanded its scope. If we recall from earlier chapters, the concept of sustainability began in the eighteenth century as a method of managing forests, and by the 1960s and 1970s it had become a reaction to industrialism and the trend toward ecological overshoot. The focus was primarily on energy consumption, our harmful economic system, population growth, and agriculture. Today, sustainability addresses these issues, in addition to many other concerns. Since 2000, for instance, the social dimensions of sustainability have gained greater attention vis-à-vis economic and environmental dimensions. Now almost anything can be viewed through the prism of sustainability—education, mental health, urban planning, gardening, architecture, and so on—to the point where it can be difficult to determine what stands outside the discourse.

Indeed, one could argue that sustainability has become an all-encompassing worldview that has supplanted and displaced democracy as the central preoccupation of the age. Whereas young people in the 1960s marched for democracy and civil rights, the youth of today are galvanized by issues of sustainability: the threats of climate change, the risks of genetically modified organisms (GMOs),[2] the need to transition away from fossil fuels, the benefits of renewables, the advantages of localism and a green economy, the desire to eradicate poverty and global inequality. Above all, sustainability, as an optimistic and hopeful discourse, has become a refuge from the cynicism of environmentalism and the apathetic malaise that grips the modern world. For many people, it seems like the only viable and socially acceptable movement to build on.

There are now a number of different models, fields, and applications of sustainability. When taken together, they demonstrate how sustainability has been integrated into various domains of human society and the economy. While presenting a comprehensive overview of the many "sites of sustainability" would be impossible, the following examples offer a glimpse of the diversity of the movement since the year 2000.

SUSTAINABILITY METRICS, METHODS, AND MEASUREMENT TOOLS

When the UN released Agenda 21 in 1992, one of the central recommendations of this sustainable development action plan was to generate "better measurement [tools]" for understanding the value of "natural capital" and the sustainability of human practices. "It is also important," the authors continued, "that national accounting procedures are not restricted to measuring the production of goods and services that are conventionally remunerated."[3] What this means is that industrial society in 1992 lacked viable methods for determining whether such-and-such a practice was sustainable. As we have seen, sustainability was essentially an unmeasurable abstraction.

Since 2000, however, sustainability has become less imprecise and more calculable as new metrics, methods, and measurement tools have developed. Older economic devices have been abandoned or supplemented by social and environmental metrics, and new sustainability assessment tools have been built from scratch.[4] The net result has been the creation of completely new kinds of information that help rate the sustainability of everything from power plants and managed forests to the apples and shrimp that we buy at the grocery store. Macro-level metrics, such as the Ecological Footprint Analysis (EFA) and the Genuine Progress Indicator (GPI), can assess the sustainability of a whole country or even the entire human population. It was virtually impossible to make these kinds of assessments before the advent of new metrics. Table 6.1 outlines several major sustainability measurement tools.

Sustainability measurement tools have become ubiquitous and culturally significant. They have allowed economists to assess the health of an economy in dynamic terms that go well beyond the expansion or contraction of economic activity and businesses to define success in terms that are not solely financial. They have promoted "green products" and allowed consumers to make more informed decisions in the marketplace. They have allowed energy analysts and ecological economists to analyze the sustainability of energy sources and the pathway to a green, low-carbon economy. They have provided architects and designers with new tools and incentives for creating sustainable, energy-efficient buildings. It has never been easier, in fact, to answer the question, "Is it sustainable?"

Table 6.1 IS IT SUSTAINABLE? A SAMPLING OF SUSTAINABILITY
MEASUREMENT TOOLS

Ecological Footprint Analysis (EFA)

This well-known tool measures human demands on nature. It is an "accounting system" that determines the "footprint" (or environmental demands) of a given political unit (city, country, etc.). It measures "how much land and water area a human population uses. . . . This includes the areas for producing the resources it consumes, the space for accommodating its buildings and roads, and the ecosystems for absorbing its waste emissions such as carbon dioxide." The system also "tracks the supply of nature." It figures out "how much biologically productive area is available to provide these services" (biocapacity). In the end, an EFA generates a numerical value that determines whether a political unit is living within its available means or is in a state of ecological overshoot. The current world footprint shows that the Earth would need the "equivalent of 1.5 planets to provide the resources [that] we use and [that] absorb our waste."[5] Many cities and countries employ EFA, as does the World Wildlife Fund for Nature, which uses it for its annual Living Planet reports.

Carbon Footprint

A cousin of EFA, the Carbon Footprint is a tool that measures the total carbon dioxide (and sometimes methane) emissions of an individual, industry, activity, or political unit. It shows, for instance, that those who live in the suburbs and commute to the city by gas-powered car have a higher carbon footprint than those who live in energy-efficient urban homes and walk or bike to work. The Carbon Footprint can be helpful in determining who (or what) is ultimately responsible for anthropogenic climate change. It can also be helpful in establishing carbon emissions reduction targets.

Life-Cycle Analysis or Assessment (LCA)

LCA is a commonly used method for assessing the complex environmental impacts associated with a product or an energy source. It is a cradle-to-grave analysis that measures all aspects of an item's history. Take blue jeans as an example. An LCA of a pair of blue jeans would measure the raw materials extracted to make the product, the processing of the materials, and the manufacturing, distribution, use, maintenance, and disposal (or recycling) of the jeans. One is then able to calculate the energy and material inputs as well as the environmental impacts of producing a pair of blue jeans. What we learn from LCA is that some products are more pollutive and energy intensive than others. This information helps establish whether an industry or product is sustainable.[6]

Energy Return on Investment (EROI)

EROI is a method used in physics, ecological economics, and the energy sector to figure out if an energy source takes more energy to produce than it ultimately yields. More precisely, it is a ratio of the amount of *usable* energy acquired from a particular energy source versus the amount of energy expended to obtain that energy source. By using EROI, one can determine if an energy source offers a net gain or net loss. EROI has been used to assess conventional energy sources (fossil fuels, nuclear) as well as renewables. Many energy analysts using EROI have concluded that corn ethanol takes more energy to produce than it ultimately yields. One of the problems with the EROI, however, is that it says nothing about the pollutiveness of an energy source, which needs to be taken into consideration when determining its sustainability. For instance, coal has a very high EROI but is extraordinarily pollutive.

continued

Table 6.1 (continued)

I = PAT

This formula was developed in the 1970s as a way of understanding human impacts on the natural environment. It assumes that impact (I) on the environment is produced by the combination of population (P), affluence (A), and technology(T) of a given society. It is still used at times but has been replaced by more nuanced metrics, such as STIRPAT.

UN Human Development Index (HDI) and the Human Poverty Index (HPI)

The HDI is an oft-criticized tool that measures the social well-being of countries around the world. It uses data on life expectancy, education, and per capita income levels to rate the relative development of a country. It is meant to go beyond the gross domestic product (GDP) as a tool for measuring the development of a country, but it lacks an ecological component and is therefore often ignored by sustainists. The HPI is meant to supplement the HDI and measures the life expectancy, literacy, and basic standard of living around the world.

UN Millennium Development Goals (MDGs)

The Millennium Project, headed by the economist Jeffrey Sachs, devised a set of MDGs that established development goals as well as measurable health and economic indicators for all 189 UN member states. The idea is to help establish a more sustainable global society by 2015, especially in the world's poorest countries, by eradicating extreme poverty, achieving universal primary education, promoting gender equality, reducing child mortality rates, combating infectious diseases (malaria, HIV/AIDS, etc.), ensuring environmental sustainability, and developing global development partnerships.

Index of Social Health

This tool measures the social well-being of a society (mainly applied to the United States). It uses a composite of 16 social indicators to determine the overall social health of a country. It is meant as an alternative to the GDP, but since it lacks ecological dimensions, it has limited applications for assessing sustainability.

Genuine Progress Indicator (GPI)

The Genuine Progress Indicator was created in 1995 by an organization called Redefining Progress. The GPI is meant to function as an alternative to the GDP. Both the GDP and the GPI are measured in monetary terms. However, the GPI doesn't measure simply the "busyness" of an economy. It takes GDP information and adjusts for factors such as income distribution, crime, pollution, long-term environmental damage, and dependence on foreign assets. It thus attempts to "redefine progress" and offers a more robust assessment of the sustainability of a society and its economy. As one might expect, a country's GPI rating often looks quite different from its GDP. The GPI shows that, in the US, "economic growth has been stagnant since the 1970s," even though the GDP has continued to climb.[7] In some places, the GPI has even declined in recent decades.

Genuine Wealth

Genuine Wealth was created by the economist Mark Anielski. As with the GPI, it is meant to function as an alternative to the GDP. It is a comprehensive accounting system, drawing on quantitative and qualitative analyses, which measures the "physical and qualitative conditions of well-being." Genuine Wealth places on the same balance sheet "human, social, natural, built, and financial capital"

continued

Table 6.1 (continued)

to gain a deeper understanding of the overall sustainability of a society and its economy.[8] The use of qualitative data is what sets Genuine Wealth apart from other metrics.

Happy Planet Index (HPI) and Gross Domestic Happiness (GDH)

The HPI, which was created in 2006 by the New Economics Foundation, and the GDH, which is a bit older, are separate metrics used to rate the average human happiness in countries around the world. They are meant as alternatives to the GDP. The idea is to measure quality of life rather than wealth or economic activity. They take into consideration ecological, social, and economic factors. In the 2012 HPI, Costa Rica ranked first in well-being even though it lacks the wealth of a large industrialized country. The United Kingdom ranked forty-first, and Japan ranked forty-fifth. The United States and Canada were not in the top 50.

Triple Bottom Line (TBL)

The TBL was created in the 1990s by John Elkington and is a form of ethical business accounting that rates three aspects of sustainability: financial, social, and environmental. It moves past simplistic and quantitative forms of accounting and rates the success of a business over and above its profitability. It seeks answers to three interlocking questions: Is the business profitable, innovative, and well managed? Is it engaging in environmentally sustainable practices? Does it aid social equality, justice, and the community? A TBL analysis can, in theory, reveal that a profitable company rates quite low in social and environmental performance. It is increasingly used in the domain of corporate social responsibility to promote corporate responsibility and a green economy.[9]

Leadership in Energy and Environmental Design (LEED Certification)

LEED certification was developed by the US Green Building Council in the 1990s and has become the most commonly used tool for measuring, rating, and promoting green building.[10] In the LEED 2009 system, a building is rated on its energy efficiency, water usage, materials, indoor environmental quality, innovation, and other attributes. A building can receive a maximum of 100 points. There are currently four levels of certification: Certified (40–49 points), Silver (50–59 points), Gold (60–79 points), and the difficult-to-obtain Platinum rating (80 points and above). Recently, LEED certification has expanded to assessing entire neighborhoods, with a program called LEED ND. LEED certification is practiced all over the world and has become the premier standard for determining the sustainability of a built structure. It also provides a major cachet for builders, owners, and/or operators, who often flaunt a building's LEED status.

The Sustainability Tracking, Assessment and Rating System (STARS)

STARS is a "transparent, self-reporting framework for colleges and universities to measure their sustainability performance."[11] It was developed via the Association for the Advancement of Sustainability in Higher Education. Numerous institutions of higher education, especially in the United States and Canada, have used STARS to gain a better sense of their environmental performance. However, the problem with this system is that it is not independently verified and requires honest and accurate self-assessments.

continued

Table 6.1 (continued)

Ecosystem-Based Fishery Management (EBFM)

This is a whole-systems approach to managing fisheries that rejects the still-dominant model of harvesting aquatic species called maximum sustained yield. Whereas maximum sustained yield is a linear system based on assumptions of a stable equilibrium, EBFM is rooted in the notion that ecosystems are difficult to predict and often change. EBFM promotes the maintenance of biodiversity and seeks a sufficient rather than a maximized yield. It also blends social and ecological data to form a holistic and sustainable approach to fishing.[12]

Ecolabels

In recent years, there has been massive growth in third-party certification of "green" products. Ecolabeling is meant to encourage sustainable consumption by informing consumers about the origin and impacts of products and services. The labels that appear on products or their packages create a visual and symbolic division between eco-friendly products and those produced by conventional and unsustainable processes.

Ecolabel: EcoLogo

EcoLogo was founded the Government of Canada in 1988 but is now widely used throughout North America. The certification program assures that the "products and services bearing the logo meet stringent standards of environmental leadership."[13] It certifies everything from cleaners and adhesives to inks and flooring materials.

Ecolabel: Certified Organic

The organic label is given to officially certified organic products, and is administered ordinarily by departments of agriculture or other bodies legally entitled to certify organics. In the United States, this is done by the US Department of Agriculture. There are different levels of organic certification, though, ranging from "100% Organic" and "Organic" (95–99% organic ingredients) to "Made with Organic Ingredients" (70–94% organic).[14] Standards can differ slightly from place to place. Organic labeling is meant to promote sustainable forms of agriculture.

Ecolabel: Fair Trade

The Fair Trade label is administered by certified national fair trade organizations. It ensures that products were produced in a socially and environmentally responsible way.[15] It is supposed to create greater equality between the developing world (the point of origin of fair trade products) and the developed world (where the products are consumed). Coffee, crafts, cotton, fruits, chocolate, and a few other products dominate the fair trade market.

Ecolabel: Food Miles

Food miles refers to the distance that a food item travels, beginning with where it is grown or processed and ending with where it is purchased and consumed. There have been some limited attempts to label food—especially fruits and vegetables—with estimated food miles, so that consumers can know where a food item originated and how much energy it took to get it to the store. More recent efforts, however, have moved away from food miles to a more comprehensive LCA of food production. The idea of food miles was developed by those who want to support local agriculture and limit the amount of long-distance foods that consumers purchase.

continued

Table 6.1 (continued)

Ecolabel: Forest Stewardship Council

The Forest Stewardship Council is an NGO founded in 1993 that certifies and labels wood, paper, and other forest products. It was created to promote the sustainable use of forests and help consumers make informed decisions in the marketplace. It has been criticized, however, for its confusion of standards and its inability to prevent clear-cutting and other unsustainable forestry practices.

Ecolabel: Marine Stewardship Council and Ocean Wise

The Marine Stewardship Council and Ocean Wise are separate certification programs meant to promote sustainable fisheries and the sustainable consumption of seafood. The Marine Stewardship Council began in 1996 and is today a large organization that sets standards and offers certified labels to qualified producers. Ocean Wise is a smaller conservation program run by the Vancouver Aquarium that offers its own label. It "works directly with restaurants, markets, food services and suppliers ensuring that they have the most current scientific information regarding and helping them make ocean-friendly decisions."[16]

The proliferation of measurement tools has forever changed our language and our cultural concerns. Terms such as "footprint," "life-cycle analysis" (LCA), "green building," and "eco-labeling" are new additions to English and indicate the emergence of a new consciousness about the impacts of human behavior. With new methods have also come new forms and fields of knowledge. It takes great expertise to certify a building as Leadership in Energy and Environmental Design (LEED) Platinum, analyze the life cycle of an automobile, or determine the energy return on investment (EROI) of a renewable energy source. Degrees and training programs have sprouted up to teach these new metrics, methods, and measurement tools, all of which are intended to provide useful information for constructing a sustainable society.

As these new forms of data have piled up and entered public discourse, they have become powerful counterweights to conventional accounting standards and the myths of progress under which industrial society has labored since the nineteenth century. That is, these new metrics create alternative forms of knowledge that challenge the status quo. They create new ways of seeing and new sets of values. LEED certification, for instance, has established an entirely novel way of assessing the achievements of architectural design. From the perspective of LEED, a beautiful and socially valuable building is one that is well designed ecologically, not just aesthetically pleasing or superficially economical. Wastefulness is now ugliness. Likewise, the GPI and Genuine Wealth offer viable alternatives to the gross domestic product (GDP). These new metrics reject conventional ideas about the health of an economy and force us to reconsider our cultural priorities and economic policies. The Triple Bottom Line (TBL) shows how cost–benefit analysis can be shallow and destructive to social and environmental well-being. It helps businesses redefine progress and shows that wealth creation—the conventional measure of success in our society—is not always beneficial for humans and the natural environment. EFA redefines the very concept of a city or a country and reveals the extent to which a human community borrows from "distant elsewheres." LCA similarly generates forms of information that simply did not exist in the past. It is now possible to understand the total energy requirements and environmental impacts of products and services, which allows producers, investors, and consumers to evaluate the sustainability of an industry.

Further, the advent of ecolabeling (organics, EcoLogo, Fair Trade, and so on) reflects the Manichean approach to consumption that divides our society—a desire for green products and sustainable consumption, on the one hand, and the inertia of destructive consumption patterns, on the other. Awareness about food miles can change consumer behavior. Organic and "conventional" foods often sit side by side in the grocery market, forcing consumers to make stark choices between opposing agricultural systems. With fair trade's challenge to "free trade," economic ideology seems to hang in the balance.

Many of these new metrics have been challenged by the defenders of the old ways on the grounds that they are biased, contain faulty assumptions, and/or privilege qualitative analysis over purely quantitative data. Even when they are accepted as valid methods, they are usually seen as "alternatives" to the norm. But despite the fact that these rival metrics remain, in most cases, secondary to and less mainstream than conventional standards, methods, and accounting practices, collectively they have made great advances in the early twenty-first century.

There are many potential examples to cite. Programs in engineering and economics commonly offer LCA and EROI training, and countless businesses, NGOs, and governments employ experts trained in these useful methods. The Marine Stewardship Council labels over 11,000 sustainable products from the world's oceans.[17] The organic industry and organic labeling have continued to grow worldwide. Even though, in 2010, organic food and beverage sales represented only 4% of total sales of foodstuffs, industry sales jumped 7.7% from 2009. The industry was worth $26.7 billion in 2010. In the same year, 11% of the fruits and vegetables sold in the United States were certified organic.[18] Despite the economic downturn in 2008, the "total amount of LEED-certified buildings grew by one billion square feet" in 2010 alone. That's a growth rate of 14%. It means that the world added 8 billion square feet of LEED-certified buildings in a single year, mainly in the United States, China, India, and the United Arab Emirates, which translates into huge reductions in CO_2 emissions—an estimated 8 million tons—since LEED buildings are energy efficient and often run on renewables. When aggregated over time, LEED-certified buildings will have prevented the release of 170 million tons of CO_2 emissions by the year 2030, which represents a not-insignificant proportion of the world's carbon footprint.[19] TBL is used by numerous small- and medium-sized businesses to drive corporate responsibility. Lastly, EFA has been employed by a range of countries, businesses, and municipalities to reduce human stresses on global ecosystems by, for instance, reducing waste, transitioning to renewable forms of energy, and emphasizing local agricultural systems.

ENERGY

There is a near consensus among sustainists that global society must run primarily on sustainable and renewable energy sources and transition away from "hard energy." What will replace fossil fuels and nuclear power? Determining *which* alternatives to use and *how* to use them remains a very vexed debate, as does the question of energy demands. Few believe that renewables could power society at current levels of consumption, and some fuels, such as corn ethanol, have proven to be very fossil-fuel reliant and yield less energy per unit than they take to produce. Other sources are still being tested, such as algae-based biodiesel and wave power. None has emerged as a panacea.

There are three main reasons why "green energy" has become a centerpiece of the sustainability movement, despite the fact that renewables remain more of a hope than a reality. First, the world is running out of fossil fuels and must look for alternative energy sources. According to Richard Heinberg, production of conventional oil, natural gas, and heavy oil all peaked around 2010. Also, extracting fossil fuels has become increasingly expensive, risky, and environmentally hazardous, as we learned from the offshore oil spill in the Gulf of Mexico in 2010. The decline of cheap and abundant fossil fuels means it is virtually impossible—not to mention undesirable—for the industrial economies of the world to continue to grow.[20] Moreover, for the vast majority of sustainists, nuclear power is seen as dangerous and unsustainable.[21] Certainly, it has never lived up to its billing as a cure-all for energy demands, which is how it was characterized in the postwar era.

Second, despite the fact that stocks of nonrenewable fossil fuels are quickly dwindling, the rate of consumption continues to grow as developing countries join the fossil fuel feeding frenzy. Our world is addicted to fossil fuels and relies on them to run the industrial economy. According to the Intergovernmental Panel on Climate Change's (IPCC) 2011 special energy report, *Renewable Energy Sources and Climate Change Mitigation*, fossil fuels remain the world's primary source of energy. Consider the statistics for 2008: Oil constituted 34.6%, coal 28.4%, gas 22.1%—that's a combined 85.1% for the fossil fuels—and nuclear 2%; renewables made up the remaining 12.9%.[22]

Third, fossil fuels are dangerously pollutive. The consensus of the scientific community is that emissions from fossil fuels are the main cause of global climate change. Of the over 36 billion tons of carbon dioxide (CO_2) emissions that enter the atmosphere every year, 81% of it comes from burned fossil fuels and the other 19% comes from deforestation.[23] The increased concentration of greenhouse gases (GHGs) has created a thermal blanket that traps heat from escaping the Earth's atmosphere, which then raises average temperatures around the globe, destabilizes weather patterns, melts ice sheets and glaciers, and produces a whole range of other negative consequences for humans, animals, plants, and ecosystems.[24]

For all three of these reasons, the challenge of the twenty-first century is to find sustainable ways to power our society, while simultaneously reducing energy consumption.

What is the status of renewable energy today? As noted above, it accounted for 12.9% of the total sum (492 exajoules) of primary energy supply in 2008. Most of that energy came from biomass (10.2%) and hydropower (2.3). Geothermal, solar, wind, biofuels, and ocean energy[25] generated a combined 0.4%. However, those numbers are higher in the domains of electricity generation, transport fuels, and heat generation. Renewables accounted for 19% of the global electricity supply in 2008. Sixteen percent came from hydro and 3% from the rest. Biofuels contributed 2% of transport fuels worldwide in 2008, although not all renewable biofuels should be considered sustainable.[26] In the same year, renewables accounted for 27% of global heating and cooling, which derived from traditional biomass sources (17%), modern biomass (8%), and solar and geothermal (combined 2%).[27] In basically all sectors of energy, though, fossil fuels remain king.

Yet the relatively low figures for renewables mask the steady expansion of the industries that produce green energy. The twenty-first century has witnessed increased government support for renewables, increased investment, declining costs, and greater public interest. In recent years, the growth rates for renewables have been stunningly high. Wind power grew 32% in 2009, which means that 38 gigawatts (GW) came online globally. Hydropower grew by 3% (31 GW added). Photovoltaic (PV) solar power connected to electrical grid systems grew an astounding 53% (7.5 GW added). According to the IPCC, "Of the approximate 300 GW of new electricity generating capacity added globally over the two-year period from 2008 to 2009, 140 GW came from renewable energy additions." The use of solar panels for heating and hot water—called "solar thermal"—grew by 21% (31 GW added). Geothermal grew by 4% (0.4 GW added). Biofuels accounted for 2% of transport fuels in 2008 and 3% in 2009, meaning that 17 billion liters were consumed worldwide in the latter year.[28]

Some countries, in fact, have made great strides in moving toward an economy that runs on renewable, sustainable, and domestically (or regionally) produced energy sources. In Denmark in 2013, nearly 30% of its electricity came from wind, and the Danish grid is connected to Scandinavian and German grids so that wind power is maximized.[29] The Danish island of Samso, which has a population of roughly 4,000, has achieved complete energy independence by using biomass, wind, and solar power.[30] In Germany, renewable energy has become a huge and profitable industry, employing hundreds of thousands of people. Thanks in large part to new onshore and offshore wind farms as well as an aggressive campaign to promote the installation of solar panels on the roofs of private residences, Germany now produces over 20% of its electricity from renewables, and that figure continues to rise.[31] Germany, indeed, has an ambitious plan to phase out its remaining nuclear power plants and decrease its reliance on coal over the next few decades. By 2050, it plans on running its society almost exclusively on renewables.

Numerous governments have become more involved in green energy. In Germany, the government used a mechanism called "feed-in tariffs" to motivate the installation of privately owned solar panels. Feed-in tariffs pay small-scale energy producers—for example, a family that has grid-connected solar panels on the roof—a fee above market rates for energy that it supplies back to the grid. That is, while the feed-in tariffs last, the average German with solar panels will turn a small profit, which then offsets the cost of the solar panels and drives interest in renewables.[32] In Ontario,

Canada, as in other parts of the world, the government established rebates for homeowners who purchased and installed solar panels. Since 2008, the Spanish government has backed the construction of concentrated solar power plants, which use mirrors to intensify sunlight and generate more electricity. The US government continues to fund and support research in algae biofuels.

For individuals, using alternative energy, especially solar, biofuels, backyard geothermal systems, and micro wind turbines, has become a cultural and political statement—a rejection of hard energy and centralized energy production—not to mention, in some cases, a rational economic decision. The use of renewables is something that can be done on the local level to encourage sustainability, and for many involved it has brought a sense of hope, freedom, and proactive engagement. However, the cost of renewables remains high (e.g., installing electricity-generating PV panels on a private residence costs at least $30,000), the fuels are only as green as the production and transmission systems that facilitate them, and they have so far failed to curb the growth in world energy consumption.

SUSTAINABLE DESIGN AND GREEN BUILDING

In the developed world, the conventionally built home is energy inefficient, reliant on centralized energy production, and frequently built with hazardous or ecologically sensitive materials. By contrast, sustainable design, which comprises planning, architecture, landscaping, and more, is an attempt to create a sustainable built environment. "Green building," one aspect of sustainable design, has been an identifiable architectural mode since the US Green Building Council created the LEED certification system in the 1990s. Since 2000, sustainable design has become more common in university programs and in communities throughout Europe, North America, and elsewhere. The efficiency standards of LEED go well beyond the minimalist efficiency standards required by most local and national governments.

Why do buildings need to become greener? First, it is often overlooked, but houses, apartments, commercial structures, and other buildings play a role in climate change. Houses in many parts of the world use oil or natural gas to power boilers, furnaces, and water heaters; and electricity often comes from coal plants and nuclear power. Globally, the building sector accounts for 8% of all end-use emissions, and in industrialized countries, especially those with cold winters, those numbers are often much higher.[33] (They're also higher if one includes the energy it takes to construct buildings.) In some urban areas of the United States, around 80% of emissions are directly tied to the building sector through heating, cooling, and maintenance.[34] The green building movement is thus motivated, in part, by the desire to reduce the pollutiveness and energy consumption of the housing sector. Second, before green building methods, little thought was given to the environmental impact of building materials. For instance, endangered tree species or lumber from unsustainably managed forests have been used to build homes in many parts of the world. Green building attempts to bring more awareness to the materials used as well as the ways in which they are sourced. Third, developers building conventional homes, especially in the suburbs, traditionally paid little attention to all those corollaries that tend to come with suburbanization and that jeopardize the ability to live sustainably: energy-intensive and pollutive carbased commuting, the destruction of forests and farmland for housing tracks, large homes that use more energy and water than smaller urban ones. Houses were treated as isolated entities without regard for community cohesion, transportation networks, or the impactful lifestyles of the people who would live in them.

What makes a building green or sustainable? In general, a sustainable building is one that runs on renewable energy sources, reduces (or eliminates) waste and pollution, uses energy and water very efficiently, is built with safe and environmentally friendly materials, is integrated into the local community and transportation networks, and promotes health, safety, and well-being. There are many different kinds of sustainable buildings, from off-the-grid Earthships that use passive solar designs, solar PV, water catchment systems, and recycled materials to straw bale homes, which use a renewable and sustainable material for walls and insulation to more conventional-looking structures that are grid-connected and use solar PV and a high-efficiency design. A building that already exists can be greened, too, through a process called "home energy retrofitting," which can involve adding new insulation to attics, installing solar panels, replacing old appliances with energy efficient ones, mounting double-paned windows, and so on.

Over the past two decades a whole series of green building assessment systems have appeared around the world. LCA is the most basic and reliable method for determining the total environmental impacts of a structure. The most well-known certification system for green buildings is the already mentioned LEED certification, which uses LCA and other techniques to assess the sustainability of a structure. The system was founded in the United States in 1993 and now has branch offices all over the world. A LEED certifier rates a building on its energy efficiency, water usage, materials, indoor environmental quality, innovativeness, and other factors. In this system, a building can receive a maximum of 100 points, with four levels of certification: certified, silver, gold, and platinum.[35] For many builders of green buildings, the goal is to construct (or retrofit) a "Net Zero" structure or something close to it, which means that it generates no annual carbon emissions and produces its own energy—or even a surplus, which is then sold back to the grid. Calculating whether a building is Net Zero is not easy, though, since many homes draw energy from the grid for part of the year (and part of the day) and supply the grid the rest of the time. Indeed, assessing the greenness of a building is a difficult task, which is why LEED certifiers need to be so well trained.

Green building and sustainable design have become booming industries since 2000. As of 2011, there was over 1.5 billion square feet of LEED-certified structures globally.[36] In 2013, the world-wide market for green construction materials was worth $116 billion, and it is projected to double by 2020.[37] The public has taken great interest in green building and energy efficiency as knowledge of climate change has grown. Obtaining LEED certification or living in a Net Zero home now has a social cachet attached to it. Also, there are now countless local contractors capable of installing solar panels or small-scale geothermal heating and cooling systems. In Germany, the thirst for home solar systems remains strong, even after the government began to scale back on feed-in tariffs. New solar power installations hit a record high in 2012, as capacity grew by more than 7.6 GW, bringing the total to over 36 GW.[38] In that year, renewables (solar, biomass, wind, and hydro) accounted for 22.9% of Germany's electricity consumption, and much of that solar power came from PV mounted on private homes, in addition to solar power plants.[39] District heating systems have also become common in Europe and North America and represent an important advance in sustainable design. These systems distribute heat, which is sourced from power plants ("cogeneration"[40]), to a grouping of commercial and/or residential buildings in an effort to reduce pollution and increase energy resourcefulness.

A common argument among sustainists is that greener lifestyles must begin at home, and indeed the home has become one of the central sites of sustainability. Builders and homeowners concerned with sustainability have responded in a range of ways, from relatively easy and inexpensive actions, such as installing energy-efficient appliances and light-emitting diode (LED) light bulbs, to more complex and expensive ones, such as installing water catchment devices, geothermal systems, solar PV, and solar thermal. According to one study, green building is well worth the efforts: Even for those who don't aim for a Net Zero home, intelligent planning and retrofitting can reduce energy consumption by 50%, carbon dioxide emissions by 39%, water usage by 40%, and solid waste by 70%.[41] It can also boost the value of a home.

URBANISM

The world is increasingly urbanized.[42] In 1900, 150 million people lived in cities. In 2000, 2.8 billion people lived in cities, which constitutes a 19-fold increase in the urban population. Since 2008, over half of the world's 7 billion people have called an urban environment home. In fact, since 1950, urbanization has outpaced the overall growth in population: the global population has grown 2.4-fold while the urban population has grown 4-fold, as more and more people have left rural areas for cities. Most of the recent urban growth has occurred in the developing world, where megacities are now a common feature. Some cities have been known to grow by 10,000 people per month.[43] Metropolitan Mexico City has 21.6 million people; Sao Paulo, 27.6 million; and Mumbai, 20.7 million. Today there are at least 25 megacities (10 million or more inhabitants) on the planet. The UN projects that the world will add about 3.3 billion more people by 2050, and much (although certainly not all) of that growth will occur in the cities of the developing world.[44]

As urban populations have skyrocketed, cities have expanded and eaten up precious farmland and natural areas. In the 10 years of the 1990s, the United States added 33 million people, mostly in cities, and as a result 2.2 million acres of land were developed *every year* in that decade.[45] With cities, of course, comes pavement. In the United States alone, an area equivalent to the arable land in the states of Ohio, Indiana, and Pennsylvania has been paved over, mainly for cars.[46] Indeed, the world's cities are now clogged with automobiles. In 1950, there were 70 million registered motor vehicles on the planet. In 2007, there were 806 million.[47] Urbanists argue that cars jam up urban areas, decrease the quality of life for urban dwellers, and contribute massively to GHGs.

Moreover, urban development has produced numerous adverse ecological consequences for humans and the planet. Many cities are dangerously overcrowded, insalubrious, and socially stratified. Housing shortages and problems with sanitation, sewage systems, and potable water affect cities throughout the developing world. Cities are also centers of consumption—of food, clothing, water, heat, fuel, and electricity. We know from the work of Rees and Wackernagel that cities stress ecosystems well beyond geographical city limits, and many, perhaps most, of the world's cities are living in ecological overshoot.[48]

As a result of these manifold problems, cities have become a key site of sustainability efforts. The first wave of urbanists who criticized traditional approaches to urban planning and so-called urban renewal—chief among them, Jane Jacobs—tended to focus on socio-economic issues, ranging from crime and slums to the need for economic diversity and livable cities. The newer approach takes a more ecological point of view and attempts to address how pollution, waste, energy consumption, and transportation relate to social and economic issues.

Perhaps the most influential movement in sustainable urbanism over the past few decades has been New Urbanism, an urban design movement that tries to reverse the trends of suburbanization, which created unwalkable, car-centered, low-density communities.

By contrast, the New Urbanists seek a return to dense, urban neighborhoods characterized by mixed-use buildings (residential, commercial, entertainment, etc.) and a vibrant sidewalk culture. They value pedestrian-only zones, abundant open spaces for leisure and social gathering, land-use strategies that prevent sprawl, bike lanes, and integrated transportation systems that reduce reliance on the automobile. Instead of tearing down warehouse districts and industrial zones, the New Urbanists stress the need to reuse existing infrastructure and bring suburbanites back into urban areas.[49] This design model has been used in many towns and cities of the United States, including Seaside, Florida, and Longmont, Colorado.

But New Urbanism is merely one approach to "green urbanism," which the urbanist Timothy Beatley defines in the following terms: planning a city that (i) exists within its ecological limits and has leaders who take responsibility for its reliance on other cities and non-urban areas; (ii) uses biomimicry to mimic the efficiency and recycling of natural systems; (iii) creates a system of circular metabolism that eliminates waste; (iv) strives for self-sufficiency, locally produced energy, locally grown food, and a strong local economy; (v) promotes social sustainability and healthy lifestyles; and (vi) emphasizes quality of life and livable communities.[50] For Beatley, whose target audience is urban planners in the United States and Canada, the best examples of sustainable cities are found in Europe, where small- and medium-sized cities tend to have much smaller ecological and carbon footprints than comparably sized cities in North America. These European cities also tend to have more pedestrian zones, higher density, more green spaces, more renewable energy, and better transportation networks, including bike lanes and light rail transit. Beatley highlights the success of municipal and regional sustainability action plans in transitioning European mid-sized cities toward sustainability. [51]

The Danish architect Jan Gehl, who is perhaps the leading urbanist of the early twenty-first century, has had some notable achievements in creating sustainable cities, which, for Gehl, means primarily urban areas that run on "green mobility"—walking, biking, and public transport. "These forms of transport," he writes, "provide marked benefits to the economy and the environment, reduce resource consumption, limit emissions, and decrease noise levels."[52] Gehl and his firm have redesigned parts of Brighton, Copenhagen, Melbourne, New York, and other cities to include more pedestrian zones, bike paths, and mixed-use buildings. Gehl is perhaps most well known for his work in New York City in 2007–2008, where he redesigned parts of Broadway, reducing car traffic, installing bike lanes, and adding pedestrian areas and public outdoor seating. In his work and his writings, Gehl always stresses the "human dimensions" of cities; if a city is not enjoyable and livable, then the urban planners have failed.[53]

But beyond Gehl and other sustainable urbanists, cities in North America, Europe, and other parts of the world have started to get serious about planning for sustainability. Countless cities now have offices of sustainability and sustainability action plans that guide local and regional policy. Many cities have adopted the Melbourne Principles, a group of 10 principles for creating sustainable urban areas. Numerous organizations promoting urban sustainability have sprouted up, from the Institute for Sustainable Communities and the Center for Resilient Cities to the massive United Cities and Local Governments. Cities such as Portland, Seattle, and Edmonton have expanded urban rail networks. Vancouver, Toronto, San Francisco, Paris, London, and others have added new bike lanes and/or bike-sharing programs. The environmental and economic logic of pedestrian zones is now widely understood in New York City and elsewhere, and car access is being limited in many downtown areas. The renaissance of local and urban agriculture represents another major indicator of urban commitments to sustainability, too, as cities look to become more self-sufficient.[54] Urban agriculture has a very long history, in fact, but food gardens and livestock were largely squeezed out of Western cities in the nineteenth century, becoming casualties of the denaturalized and sanitized modern city.

Despite these achievements, notable problems still remain. Cities all over the world continue to grow and consume. Suburbs continue to encroach on natural areas. Most urban planners remain pro-growth, and "growth management" plans have often been little more than greenwashed strategies for increased urban growth. The scholar of urbanism Gabor Zovanyi argues that, in the United States, one would be "hard pressed to identify any local jurisdictions that have implemented absolute caps on future growth."[55] The community planning consultant Eben Fodor has debunked many of the pro-growth arguments for cities, showing that expanding a city does *not* solve unemployment problems, bring in higher net taxes, or make housing any more affordable.[56] And yet some planners still think growth is the answer.

Finally, revitalizing urban areas has often created problems with gentrification, as property values suddenly rise and squeeze out underprivileged local inhabitants. Many cities with strong commitments to sustainability (Seattle, San Francisco, Boulder) are dominated by what the urban theorist Richard Florida calls the "creative class"—well-educated people whose jobs and activities require independence and creativity.[57] Yet in cities such as Hamburg, Germany, where the city council drew explicitly on Florida as inspiration for its sustainability plans, there has been strong resistance to what is seen as yet more gentrification. For activists in Hamburg, a "creative class city" is a merely a euphemism for exclusion, social inequality, growth, and ecological apathy.[58]

TRANSPORTATION

The creation of sustainable transportation systems is often a re-action to the ubiquity and destructiveness of the automobile. As noted above, the world now has over 800 million motor vehicles on the road, and that number continues to grow as the developing world scrambles to replicate the developed world's transportation mistakes.

Why does the world need greener transportation systems? Simply put, because our car-centric, fossil-fueled system is unsustainable. Globally, the transport sector accounts for 13% of all GHGs—that figure is higher if you include industrial transportation—since most motor vehicles and nearly all airplanes run on pollutive fossil fuels.[59] Even if the automobile were to run on environmentally friendly fuels, which it could, sustainists argue that there would still be good reason to limit its use. Indeed, the automobile has had a plethora of negative impacts on society: it allowed for the creation of the suburbs, led to the paving over of a not-insignificant percentage of the Earth's land, and created new demands for fossil fuels. By the year 2000, in the United States alone, cars had killed or maimed over 250 million people, consumed over 8 million barrels of oil *every day*, created dependencies on foreign oil, required $200 million in maintenance costs *per day*, killed a million wild animals *per week*, created noise, exacerbated health problems (asthma, emphysema, and so on), emitted enormous quantities of GHGs, and generated 7 billion pounds of unrecycled scrap and waste *every year*.[60] A gallon of gas that is burned in a combustion engine emits about 19 pounds of CO_2, and an additional 9 pounds are emitted in the production and processing of the fuel.[61] When you add to the mix the fact that fossil fuel stocks are declining, as consumption rates rise, it's easy to see that the world cannot continue to rely on the gas-powered automobile.

What, then, does sustainable transportation look like? As Gehl puts it, "green mobility" emphasizes walking, biking, and public transportation, all of which decrease overall reliance on fossil fuels and obviate the need for more roads and highways. Gehl has shown that limiting the use of cars in urban areas substantially improves the quality of life of local residents.[62] The renewed interest in bike lanes, pedestrian zones, light rail trains, integrated mass transit systems, and green corridors in Europe and North America suggests that sustainable transportation has become more socially valued. Another indication of changing values is the "live by your workplace" movement, touted by sustainists and city planners.

But what about outside of and between cities? Although cars and airplanes still dominate long-distance travel, there is an increasing interest in high-speed trains. Germany, France, China, Japan, and the state of California have built high-speed trains in recent years. Yet trains and all other forms of mechanized transportation are only as green as the fuels that power them. To be sustainable, trains need "green" sources of electricity; airplanes need to run energy sources other than jet diesel. To this end, great efforts are currently under way to expand the production of second-generation biofuels made from algae and other micro-organisms, which can replace automobile and jet diesel. Algae-based transport fuels have attracted attention from the US military, Boeing, Exxon Mobil, and an array of airline companies, including Virgin Atlantic, which has run test flights of biodiesel-diesel blends.[63] In the end, sustainable transportation is dependent upon both green energy and intelligent planning.

HIGHER EDUCATION AND RESEARCH

Since 2000, the sustainability movement has had increased visibility within institutions of higher education throughout Europe and North America, in terms of both administrative action and scholarly research. In 2013, the Association for the Advancement of Sustainability in Higher Education (AASHE) counted 862 schools in Canada and the United States as members. Many of those colleges and universities have offices of sustainability, which craft and coordinate campus sustainability policies. AASHE also manages the Sustainability Tracking, Assessment and Rating System (STARS), which is a "transparent, self-reporting framework for colleges and universities to measure their sustainability performance." Over 250 AASHE members have used STARS to reduce their waste, carbon footprint, and consumption of electricity and water.[64] Furthermore, chancellors and presidents have committed to promoting sustainability in higher education. Over 350 universities worldwide in 40-plus countries have signed the Talloires Declaration (1990), for instance, which is a "ten-point action plan committing institutions to sustainability and environmental literacy in teaching and practicing."[65]

Sustainability is a growing subject of academic research, too. Engineering, arts, architecture, agriculture, urban planning, politics, business, economics, sociology, and many other academic fields are involved in studying and assessing sustainability. Since 2000 there has been a rapid development of university programs that offer degrees in sustainability, from the pathbreaking College of Sustainability at Dalhousie University, to the School of Regional and Community Planning at the University of British Columbia, to the PhD program in sustainable development at Columbia University. The latter program "combines a traditional graduate education in the social sciences, particularly economics, with study in the natural sciences and engineering, to prepare scholars who are uniquely situated to undertake serious research and policy assessments in furthering the goal of sustainable development."[66] Indeed, there are now several hundred programs around the globe that deal with some facet of sustainability. AASHE lists 1,377 sustainability-focused academic programs. Even universities that lack programs in sustainability often have courses in the subject in various fields and certificates in sustainability.

The number of journals dedicated to sustainability has expanded, too, which has provided an outlet and forum for new sustainability-related research. *Ecological Economics* has been around since the 1980s, but there are many new specialized journals: *International Journal of Sustainability in Higher Education, International Journal of Sustainable Development, Journal of Sustainability and Green Business,* and many others. These journals have given shape to the numerous academic subfields of sustainability.

Despite the growth of sustainability in the research interests, curriculum, and administration of universities, important questions still remain about the ability of higher education to promote green values. Rees, for instance, has suggested that colleges and universities often "impede sustainability" by remaining wedded to fields and methods that reproduce the habits and beliefs that generated unsustainability.[67] For example, most business programs still minimize social and environmental responsibility and teach that profitability is the sole purpose of a firm. Most economics departments are still dominated by neoclassical economists who uncritically encourage growth and see the GDP as a useful measure of economic health. As a result, sustainability has often found a home in newer and less traditional academic fields—art and design, resource economics, environmental sociology, renewable resources, international and public affairs, and so on, and because of this, there is often conflict on university campuses between sustainists, ecologists, and ecological economists, on the one hand, and those tied to business as usual, on the other. It can get complicated for students, who are unknowingly exposed to conflicting bodies of knowledge. But despite Rees' concerns, it's clear that universities have played and will continue to play an important role in the development of the sustainability movement.

THE ECONOMICS OF SUSTAINABILITY AND THE GREEN ECONOMY

Ecological economics forms the foundation of the economics of sustainability, which is a growing school of economic thought that assesses the unsustainability of today's world and advocates for the creation of a green economy. This alternative brand of capitalist economics builds off the work of Mishan, Daly, Schumacher, the Club of Rome, and the other first-wave ecological economists of the 1970s. It rejects neoclassical economics and the notion that the ultimate purpose of an economy is endless growth. It seeks a steady-state economy or something akin to it. It harmonizes economic and environmental values. It uses new methods to create innovative forms of knowledge. It favors regulation, the preservation of natural capital, social equality, the end of waste and pollution, and the abandonment of reckless consumerism. The economists associated with the economics of sustainability include William Rees, Mathis Wackernagel, Tim Jackson, Peter Victor, Richard Heinberg, Clive Spash, Joshua Farley, and Mark Anielski. They have produced a number of works with titles such as *Prosperity Without Growth* (Jackson), *Managing Without Growth* (Victor), and *The End of Growth* (Heinberg).

Although this economic school has not succeeded in displacing neoclassical economics in university courses, government ministries, or banks, the field has enjoyed substantial development and increased social legitimacy since 2000. Ecological economics is certainly taken more seriously today than it was in the 1970s. This shift in attitude is due in large part to public acceptance of the world's troubling social, environmental, and economic problems. It is now generally acknowledged that peak oil production has already occurred. Anthropogenic climate change is a reality acknowledged by scientists and the majority of the population.

The Great Recession of 2007–2009 brought the world's attention to the pitfalls of the current economic paradigm. Social inequality has grown globally while aggregate happiness in much of the industrialized world has plateaued or declined. The economics of sustainability offers potential solutions to all of these problems, and thus it has gained the attention of the public, NGOs, businesses, and many governments. It is no surprise that Tim Jackson held a well-publicized TED Talk in 2010 on life beyond growth and consumerism.

What is the message of these sustainability economists? First, they all deny the developed world's need for more economic growth and instead argue for an economic system that respects stability and ecological limits. Broadly, they agree with the Club of Rome's assessment about the limits to growth. They believe that the Earth, and its human civilization, cannot cope with the pollution generated by an industrialized economy, nor can it deal with the ecosystem destruction and food requirements of a mounting global population. As Jackson puts it, "This extraordinary ramping up of global economic activity [since 1950] has no historical precedent. It's totally at odds with our scientific knowledge of the finite resource base and the fragile ecology on which we depend for survival."[68] Robert and Edward Skidelsky write in the same vein: "We agree that, for the affluent world, growth is no longer a sensible goal of long-term policy."[69] Peter Victor agrees: "Rich countries can manage without growth."[70] Richard Heinberg is the most emphatic of the bunch:

> But as the era of cheap, abundant fossil fuels comes to an end, our assumptions about continued expansion are being shaken to their core. The end of growth is a very big deal indeed. It means the end of an era, and of our current ways of organizing economies, politics, and daily life.

I am asserting that *real, aggregate, averaged* growth is essentially finished, though we may still see an occasional quarter or year of GDP growth relative to the previous quarter or year, and will still see residual growth in some nations or regions.[71]

Second, sustainability economists reject the metrics and methods used in conventional economic analysis. They criticize the GDP/GNP and the notion that "more is better." They lament the idea of "externalizing" environmental and social costs. They also reject the use of taxes and subsidies to support fossil fuels, cars, consumption, and growth.

Third, they favor an economic system that values happiness, redefines progress as something other than growth, and equates "prosperity" with fulfillment, personal development, and social justice.[72] Of course, ideas about what constitutes the "good life" transcend economics; there is a general consensus that the developed world needs to discard materialistic cultural values and instead seek "inward richness." But this message relates directly to economics, too, in that sustainability economists support macroeconomic policies that nurture "qualitative development" over the accumulation of more stuff—and development that ultimately results in the fair distribution of wealth and resources and the ability of humans to flourish and live dignified lives.

The ultimate goal for all of these economists is a "green economy": one that is low carbon, democratically decentralized and environmentally sustainable and that promotes equality, well-being, and life satisfaction.[73] Nicholas Stern, in his famous *Stern Review on the Economics of Climate Change*, suggested in 2006 that it would cost 1% of global GDP per annum to transition away from fossil fuels and create a green economy, although in 2008 he doubled that estimate to 2%, to account for faster-than-expected climate change.[74] Backers of a green economy also tend to reject growth for growth's sake, as we saw.[75] Indeed, ecological economists argue that the creation of a green economy necessitates a revolution *within* capitalism. Jackson wants a "less capitalistic" capitalism, and Rifkin seeks a new kind of collaborative and "distributed capitalism."[76]

What do sustainability economists propose as alternatives to neoclassical economics and business as usual? Since the 1990s, sustainability economists have either invented new economic metrics or incorporated methods from ecology, physics, and engineering into older economic models. The result has been the creation of new forms of knowledge that support economic sustainability. It is here where we see the greatest difference between the economics of sustainability and the first wave of ecological economics. The economists of the 1960s and 1970s put forth important critiques of neoclassical economics and industrialism, but created relatively little in the way of sustainability measurement tools or constructive alternatives to a growth-based economic system.

By contrast, sustainability economists in recent years have created or utilized a whole range of tools, methods, and economic metrics that value sustainability. As noted above, Rees and Wackernagel created EFA as a way of understanding how economies rely on ecosystem services and often live beyond their available ecological means. The aforementioned LCA is another tremendously useful tool, used by sustainability economists to assess the complex environmental impacts of a product or an energy source. It reveals all the hidden costs of an industrial economy. Before the advent of LCA, it was impossible to calculate the energy intensiveness and pollutiveness of an industry or a product. The GDP, which measures the "busyness" of an economy, has been reformulated or set aside for new measures of economic well-being. The GPI, which was created in 1995 by a group called Redefining Progress, takes GDP data and adjusts for social and environmental factors, such as crime, pollution, and income distribution. It shows that industrialized countries have not been growing economically for some time. The United States, for example, has been stagnant since the 1970s. Genuine Wealth, another alternative to the GDP, uses quantitative and qualitative analyses to measure aggregate well-being in a society. It places on the same balance sheet "human, social, natural, built, and financial capital."

The study of aggregated national happiness has become a thriving subfield within economics and has emerged as one of the most powerful counterweights to the GDP. Whereas the GDP is a quantitative metric, national happiness is a qualitative measure of social well-being and life satisfaction that takes into consideration ecological factors. Interest in aggregate happiness traces back to Richard Easterlin's pathbreaking 1974 article, "Does Economic Growth Improve the Human Lot? Some Empirical Evidence," and has been studied by Amartya Sen, Mark Anielski, Richard Layard, the New Economics Foundation, and others.[77] The two most common metrics that rate communal happiness are the Happy Planet Index (HPI), which was created in 2006, and an older (and often revised) measure called Gross Domestic Happiness. The HPI produces periodic ratings which make for fascinating reading. As one might expect, the richest countries often fare poorly in life satisfaction. In the 2012 HPI, Costa Rica ranked first even though it lacks the wealth of a large industrialized country. The United Kingdom ranked forty-first, and Japan ranked forty-fifth. The United States and Canada did not make the top 50.

The HPI seems to indicate a kind of sweet spot in terms of economic development, beyond which a society becomes merely corpulent and unhappy. Costa Rica, the Bahamas, and Brunei all have high Life Satisfaction ratings but a per capita GDP below $20,000.[78] The upshot is that severe poverty generally makes people unhappy, whereas growing wealth has diminishing returns on life satisfaction. There is a middle ground in which the citizens of a country are very happy without being rich or consumer-oriented. The analysis of communal happiness has, at times, received strong criticism for the methods that it employs: How does one aggregate joy? What if one country's idea of happiness differs from another's? Nonetheless, it remains an important method for challenging the assumption that consuming more stuff makes people happier.[79]

Sustainability economists also reject the concept of "externalities," which, as we saw in chapter 4, exclude from economic analyses the social and environmental costs associated with industrial production. What they seek, instead, is a more inclusive "true cost economics." Rees offers a detailed explanation:

Neoclassical economists have traditionally been content to allow the prices of goods and services to be determined solely by the law of supply and demand. However, in unregulated markets, only direct producer costs (for rent, labor, resources, and capital, for example) are reflected in consumer prices. The prevailing cost-price system does not account for the collateral damage to ecosystems, human communities, or population health caused by many production processes. These external (outside the market) costs are born disproportionately by third parties or society at large—and, of course, the ecosphere. Because negative externalities represent real costs, the goods and services inflicting them enter the marketplace at prices below their true cost of production. Such underpricing leads to over-consumption, inefficient resource use, and pollution—all classic symptoms of market failure.

By contrast, in a true cost economic system, consumer prices would incorporate environmental, health, and other welfare damage costs of production. When prices "tell the truth" about costs, consumers adjust their consumption patterns accordingly, purchasing fewer ecologically costly goods. Markets would operate more efficiently, producers would innovate and adopt cleaner production processes, total production/consumption would decline (a good thing in a resource-stressed world), pollution and health costs would be reduced to insignificance, and third parties would be relieved of an unfair burden.[80]

True cost economics thus addresses market failures and allows consumers to grasp the *real* cost of goods and services. True cost accounting reveals that ecosystem services are far more valuable (and expensive) than people generally realize.

Sustainability economists also tend to favor greater government regulation of markets and a range of extra-market mechanisms. As Rees notes, "Correcting for market failure requires government intervention."[81] They contend that free trade and market deregulation causes overconsumption, resource depletion, loss of biodiversity, social inequality, and pollution. They have developed a number of regulatory correctives in response. The first are pollution taxes, eco-taxes, and other so-called Pigovian taxes that make pollution and ecological destruction costlier and thus less attractive. The idea is that there should be a green tax shift that "punishes" (via taxes) the things that we don't want—GHG emissions, hazardous waste, and so on—and "rewards" (via the absence or reduction of taxes) the things that we do want—namely, sustainable practices. A second corrective is the practice of shifting government subsidies from pollutive to green industries. One estimate shows that global fossil fuel subsidies in 2012 reached $775 billion.[82] Those subsidies not only create the false impression that fossil fuels are cheap and abundant but also show the extent to which global cultures value and rely on hard energy. As cultural values shift toward the green economy, the idea is that subsidies should go toward industries that develop sustainability.

Third, some of these economists support cap-and-trade systems that impose upon industries fixed emissions levels with tradable pollution rights. Cap-and-trade schemes can, in theory, maintain a market system while encouraging emissions reductions and shifts to greener technologies. Finally, there is a consensus among sustainability economists that natural capital—the goods and services that come from the natural environment—need to be protected, restored, and valued. In our current economic system, natural capital gains value only when it's put into productive use. However, an economy that seeks sustainable reliance on nature's offerings must value natural capital before it is "used" in the economy. A tree standing in a forest is sucking carbon dioxide emissions out of the air—yet neoclassical economics does not place any inherent value on that crucial ecosystem service.

The economics of sustainability has never been in a better position to challenge the dogmas of neoclassical economics. An increasing number of universities, including the University of Edinburgh and the London School of Economics, offer MA programs in ecological economics. New research on the pitfalls of growth, our addiction to finite resources, and the lack of life satisfaction in the industrialized world has made its way into university courses and public consciousness. The metrics used in this field, most prominently the EFA, have become mainstream and legitimate ways of challenging eco-apathy and the growth paradigm. The advantage of this emerging field of economics is that it's conventional enough to have broad appeal—after all, it's still a form of capitalism—while also recalibrating economic thought around green values, life satisfaction (vs. materialism and growth), and the realities of ecological limits.

BUSINESS AND FINANCE

The business world is increasingly interested in sustainability, not only because such interest is seen as a way of keeping up with the times but also because sustainable business practices can allow firms to profit and function over the long term. After all, cutting waste and increasing energy efficiency saves money, and a growing number of consumers want to buy goods and services that are genuinely green. Employee satisfaction is part of business sustainability, too, and many companies have sought to increase the well-being of employees, thereby reducing turnover and increasing employee loyalty. As with universities and city governments, many businesses now have departments or offices of sustainability that coordinate the corporate social and environmental policies. Others have turned to green consulting agencies to reduce waste and carbon footprints. For instance, in 2005 Walmart hired the Rocky Mountain Institute in an effort to double the fuel efficiency of its trucking fleet by 2015.[83] More recently, Shell Canada hired Green Analytics, a Canadian consulting firm, to help the oil giants avoid sensitive ecological areas and reduce its industrial footprint in the Albertan oil sands.[84]

Is corporate interest in sustainability mere window dressing? Won't corporations do anything to turn a profit? This is an ongoing and rather vexed debate. Certainly, some corporations have been guilty of setting up bogus "environmental" research institutes which serve as cover for conventional extraction, production, and land use practices, and many corporations have been accused of greenwashing—of twisting language or using dubious ad campaigns in an effort to cast themselves as green. Weyerhaeuser, Chevron, Monsanto, and many other corporations have been accused of having a false commitment to the values of sustainability.[85] No doubt many of these accusations are legitimate, but it's also the case that some corporations have a more genuine interest in operating sustainably. Since 2000, corporate social responsibility (CSR) has become a powerful movement within business ethics, replete with its own journals, seminars, gurus, and manuals. CSR is an industry-generated attempt to improve environmental and social accounting and accountability.

TBL, as mentioned in the previous chapter, is by far the most recognizable sustainability accounting method used by firms. It measures the success of a company based on its track record for generating profit, promoting social well-being, and operating in an environmentally responsible way. The motto "people, planet, profit" is now commonly heard in the business world. Although relatively few multinational corporations have shifted completely to TBL accounting, many smaller ones have taken up the challenge. Companies with a stake in sustainability seem to have the easiest time adopting TBL standards. For instance, Southern Energy Management, a small American corporation that installs solar panel systems, uses TBL because it is consistent with its overall mission.[86] It's much more difficult for an oil company or mega-retailer to use TBL, because it would likely require that the corporation make drastic changes to its business practices or even rethink its raison d'être. At the end of the day, Walmart sells imported plastic widgets, and Shell extracts fossil fuels from the ground. Only a major shift in orientation could make those corporations sustainable, no matter how one defines the term.

Yet according to the American attorney and environmental advocate Van Jones, the "green-collar economy" has grown dramatically in recent years as businesses have capitalized on newfound interest in green services and products and as governments have dedicated more funds for renewables and green infrastructure. Jones defines green-collar jobs as "family-supporting, career-track, vocational, or trade-level employment in environmentally-friendly fields." He gives as examples "electricians who install solar panels; plumbers who install solar water heaters; farmers engaged in organic agriculture and some biofuel production; and construction workers who build energy-efficient green buildings, wind power farms, solar farms, and wave energy farms." One might add to that list technicians who build and maintain light rail systems, civic employees who build bike lanes, and even those working for green consulting agencies. Jones shows the profitability of the green-collar industry, too. In 2006, in the United States alone, "renewable energy and energy-efficiency technologies generated 8.5 million new jobs, nearly $970 billion in revenue and more than $100 billion in industry profits." [87] In Germany, the renewable energy sector alone employed nearly 370,000 people in 2010.[88] Thus, there is a clear motivation for businesses to not only green their own practices but also to get more heavily involved in the emerging green economy.

The world of green finance has grown, too, since 2000. One of the first banks to rethink finance was the Grameen Bank, a Bangladesh-based lending institution that offers microcredit (small loans) without collateral to the poor of South Asia. About 98% of loans are given to women, often in small groups, with the idea that small loans can stimulate the local economy and encourage sustainable self-sufficiency. In 2006, the bank and its founder, Muhammad Yunus, were awarded the Nobel Peace Prize.[89] The Grameen Bank has become a model and inspiration for new forms of credit and a business-based approach to sustainable development. For instance, nonprofits such as Kiva and E + Co have become leaders in providing loans (often micro loans) to and investing in green, community-supporting industries in the developing world.[90] In recent years, Internet-based crowd financing has become a growing industry (although clearly not all of the initiatives supported through crowdsourcing aim to be green). Indeed, a whole range of alternative financial mechanisms have appeared in recent years: self-invested personal pensions, local community-owned energy companies (such as the ones that exist in Denmark), community shares and bonds, revolving funds, social investing funds, and so on.[91]

It's clear to many sustainists and financiers that older models of capitalist finance need to be rethought in light of the demands of the green economy.[92] To this end, the former managing director of JPMorgan, John Fullerton, created the Capital Institute in 2010 as a means of creating "an economic transition to a more just, regenerative, and sustainable way of living." Fullerton asserts that "our finance-driven economic system is in urgent need of a new story, with a new roadmap."[93] A new approach to investing in clean tech, for instance, is important for the sustainability movement because energy businesses require a lot of capital and yet rarely offer quick returns. In fact, the emphasis on quick returns has led to a recent downturn in green energy investment. According to the New York Times, "Worldwide in 2012, venture capital investing in clean technologies fell by almost one-fourth, to $7.4 billion, from 9.61 billion in 2011.... In other words, clean-energy companies can't rely only on the classic venture-capital approach in which investors demand a fat, fast return."[94] For the time being, government grants, corporate partnerships, and "a willingness to pursue higher-value product lines en route to entering larger, but lower-margin markets" is what is keeping many clean tech firms afloat. Clearly, a green economy needs different forms of lending and investment than those of a traditional capitalist economy—ones that make sense for the new realities of green energy production.

The greening of business and finance is an ongoing process. Consulting firms such as the Rocky Mountain Institute, the Natural Step, and many smaller agencies remain active suppliers of ecological wisdom to the world's corporations. Green products, from local organics to fair trade goods, continue to gobble up larger shares of their respective markets. Although investment in clean tech stalled in 2012, many domains of the green economy continue to expand, and green-collar jobs are more abundant today than ever. But the transition to a green economy requires that more corporations get on board with sustainability and willingly transform their own practices. The business world can only become truly sustainable if it redefines its own mission. As Hawken wrote back in the 1990s, "The ultimate purpose of business is not, or should not be, simply to make money. Nor is it merely a system of making and selling things. The promise of business is to increase the general well-being of humankind through service, a creative invention and ethical philosophy."[95] CSR and TBL have initiated the process, but enormous challenges remain, especially among those corporations that have such large stakes in the unsustainable industrial practices that helped to create our ecological crisis.

SOCIAL SUSTAINABILITY: EQUALITY, DEMOCRACY, SOCIAL JUSTICE, WELL-BEING, AND POVERTY ERADICATION

Whereas the environmentalists and ecological economists of the 1960s and 1970s focused on the environmental and economic dimensions of living sustainably, there has been increased attention since 2000 on the social dimensions of sustainability—that is, on the ways in which equality, democracy, social justice, well-being, and poverty (especially in the developing world) relate to economics and the environment. Sustainists contend that a sustainable society is one that not only preserves natural capital, eliminates waste, and establishes economic stability but also promotes human happiness, equality, and well-being. This emphasis on human welfare is one of the central features that separates classic environmentalism from the sustainability movement. For sustainists, faulty economic systems and environmental destruction are always linked to social problems.

What is "social sustainability"? For Richard E. Stren and Mario Polèse it is defined as "policies and institutions that have the overall effect of integrating diverse groups and cultural practices in a just and equitable fashion."[96] Jonathan M. Harris and Neva R. Goodwin offer a similar definition: "The social dimension [of sustainability] may be defined as progress toward enabling all human beings to satisfy their essential needs, to achieve a reasonable level of comfort, to live lives of meaning and interest, and to share fairly in opportunities for health and education."[97] By contrast, a socially unsustainable society is characterized by extreme poverty and/or an inability for citizens to live safe, meaningful, and satisfying lives. Even if an unjust and unhappy society somehow managed to live within its ecological limits, social unsustainability would inevitably lead to socio-political unrest and even revolution. A stable society therefore requires justice, equality, democratic institutions, and the conditions for life satisfaction. For Harris and Goodwin, "a socially sustainable system must achieve fairness in distribution and opportunity, adequate provision for social services, including health and education, gender equity, and political accountability and participation."[98] Gehl notes that "social sustainability also has a significant democratic dimension that prioritizes equal access to meet 'others' in public spaces."[99]

Much of the focus of social sustainability has been on development and poverty eradication in the developing world. Sustainable development is currently being rethought and reoriented in a way that balances the three Es, so that development benefits the poor, works with local needs and circumstances, and promotes equality while also encouraging environmentally responsible practices.[100] In the past, many top–down development initiatives intended to help poorer societies have made little sense either socially or environmentally. There are many examples to cite, such as the International Monetary Fund's (IMF) role in exacerbating the food crisis in Malawi, in which "IMF-led economic liberalisation . . . undermined farmers' access to vital agricultural inputs and eliminated consumer subsidies and food price stabilisation," or USAID's disastrous attempt to establish apple farming in inhospitable rural Nepal.[101]

Daly has argued that the central problem with the development policies of the UN, IMF, and the World Bank in the 1990s was that they reproduced in poorer countries the same economic and industrial mistakes made by the industrialized world. Instead of focusing on growth and development in the developing world, the focus was always on how industrialized countries could grow and profit from development elsewhere. Moreover, local circumstances and needs were often ignored by distant bureaucrats with ideas that looked good on paper.[102] As a result, according to Harris and Goodwin, "the benefits of development have been distributed unevenly, with income inequalities remaining persistent and sometimes increasing over time." Moreover,

> there have been major negative impacts of development on the environment and on existing social structures. Many traditional societies have been devastated by overexploitation of forests, water systems, and fisheries. Urban areas in developing countries commonly suffer from severe pollution and inadequate transportation, water, and sewer infrastructure. Environmental damage, if unchecked, may undermine the achievements of development and even lead to collapse of essential ecosystems.[103]

David Griggs and his colleagues formulated in 2013 a new set of sustainable development goals that balance socio-economic and environmental concerns. The goals are to have "thriving lives and livelihoods, sustainable food security, sustainable water security, universal clean energy, healthy and productive ecosystems, and governance for sustainable societies."[104]

One of the most ambitious attempts to implement a balanced policy of sustainable development is the UN's Millennium Development Goals (MDGs). Established in 2000, the MDGs are eight development goals that are supposed to be met worldwide by 2015. All of the UN member states agreed to meet these goals, which are as follows:

1. Eradicate extreme poverty and hunger;
2. Achieve universal primary education;
3. Promote gender equality and empower women;
4. Reduce child mortality rates;
5. Improve maternal health;
6. Combat HIV/AIDS, malaria, and other diseases;
7. Ensure environmental sustainability; and
8. Develop a global partnership for development.[105]

To date, progress has been rather uneven in the different countries involved in the MDGs.[106] One part of the plan was to have the world's wealthiest countries commit to donating 0.7% of their GDP to aid and development, but that target has not been met in most countries. The Millennium Villages Project, an offshoot of the MDGs that is run by the American economist Jeffrey Sachs, who also ran the Millennium Project at the UN in the early 2000s, focuses more narrowly on Africa. It is a program that is meant to boost agricultural yields, control malaria, create health care services, improve water supplies, and expand education. According to Sachs, at least, the Millennium Villages Project has made measurable progress toward meeting its goals, including reduction of severe poverty.[107] Malaria prevention is another apparent success story: "The global estimated incidence of malaria has decreased by

17 per cent since 2000, and malaria-specific mortality rates by 25 per cent," which means that "1.1 million deaths from malaria were averted."[108]

Does poverty cause environmental destruction and political unrest, or does environmental destruction and political unrest cause poverty? This is an ongoing and often circular debate. But it's clear that poverty, political turmoil, and environmental troubles tend to go hand in hand. Diamond, among others, has argued that the modern world's perennial political and environmental trouble spots are identical.[109] He also shows the contrast in forestry practices between Haiti—the Western Hemisphere's poorest country—and its less impoverished neighbor, the Dominican Republic. "Today, 28% of the Dominican Republic is still forested, but only 1% of Haiti is."[110] For the new generation of sustainists involved in sustainable development, the goal is to end poverty, not only to improve the lives of the impoverished but also because the world's poorest countries, such as Haiti, tend to destroy local environments out of sheer desperation. The goal is to develop sustainably without setting poor countries on the same track of overconsumption and unchecked emissions on which the industrialized world is now trapped. For Daly, "sustainable development . . . necessarily means a radical shift from a growth economy and all it entails to a steady-state economy, certainly in the North, and eventually in the South as well."[111]

One of the greatest challenges for social sustainability in the industrialized world is to create opportunities for all social classes to get involved in green living. In the words of Van Jones, there needs to be "eco-equity" so that the poor and the oppressed are given the opportunity to live sustainably and find work in the growing green economy. "If the green economy remains a niche market, even a large one, then the excluded 80 percent will inevitably and perhaps unknowingly undo all the positive ecological impacts of the green 20 percent."[112] Jones is undoubtedly correct to argue that the costliness of living sustainably has blocked the involvement of the urban poor in many aspects of the sustainability movement. After all, organic foods tend to be more expensive than conventional foods and installing solar PV systems is cost prohibitive for the poverty stricken. Furthermore, for the poor, the difficulties of everyday life can make green values seem foreign and irrelevant. Even though social sustainability has many facets—not the least of which is dealing with poverty and social inequality in the developed world—closing the gap between the eco-privileged and the eco-underprivileged remains an important stumbling block to sustainability. There is increasing agreement that the sustainability movement needs to be inclusive to everyone and educate all citizens in the joys and benefits of living sustainably.

Finally, recent efforts have been made to associate sustainability with human health and well-being and to better understand the health risks of living unsustainably. A socially sustainable society is one that ensures the safety and well-being of all citizens. Unsustainability adversely affects well-being in numerous ways. For starters, pollution has myriad negative consequences on human bodies. Fossil fuel emissions have led to a "spectacular increase" in ailments such as asthma, emphysema, heart disease, and bronchial infections."[113] Toxins and pollutants affect cellular biology and can cause cancer, which is why the cancer researcher Karl-Henrik Robèrt first became interested in the sustainability movement in the 1980s.[114] A suburban, car-centric lifestyle, combined with low access to healthy foods, has also contributed to obesity rates. Overcrowded and noisy cities drive up stress and anxiety levels, while excessive consumption in affluent countries, according to happiness researchers, tends, paradoxically, to drive down life satisfaction. In theory, a socially sustainable society is one in which citizens are content to live modest lives and value satisfaction and health over money and consumption.

FOOD, LOCALISM, AND COMMUNITY SELF-SUFFICIENCY

Gardening, agriculture, and the local economy have become crucial sites of sustainability. With increased awareness about reliance on pollutive (and often foreign) fossil fuels, combined with a growing realization that industrial agricultural processes are harmful to soils, waterways, plants, animals, and human bodies, there has been a tremendous renaissance since the 1990s in producing and consuming locally and sustainably produced foods and materials. In fact, interest in eating local, healthy, organic, non-GMO foods has now gone mainstream. The desire for localism has been called a renaissance since virtually all humans in pre-industrial societies once subsisted almost exclusively off of locally produced goods. As with so many other aspects of the sustainability movement, the shift back to local foods and the local economy is a kind of updated return to origins. It's also a conscious powering down and a rejection of globalization.

The local food movement has grown rapidly since 2000. Much of this newfound interest in eating locally grown (or ranched) foods stems from the organic movement, which took shape in the 1970s and revived pre-industrial farming practices based on natural fertilizers and pesticides. Eating locally also gained popularity via permaculture, which, as we have seen, was developed in Australia in the 1970s, and stressed the advantages of sustainable, self-sufficient, and high-yield gardening. Local, often organic, food is now widely available in markets throughout the Western world. Urban farming is practiced in many cities and food co-ops are now common. The number of farmers' markets has grown tremendously in recent decades, too. The US Department of Agriculture (USDA) defines "local food" as any agricultural food product that is purchased within 400 miles of where it was produced (or within the state in which it was produced). Others have tried to shrink the definition of local. Alisa Smith and J. B. MacKinnon published a best-selling book in 2007 called *The 100-Mile Diet*, which has become the bible of the local food movement.[115]

According to the USDA, the local food market remains small compared to the international and long-distance food market, but it has grown remarkably in recent years. In the United States, the direct-to-consumer local food market was worth $1.2 billion in 2007, compared with $551 million in 1997. The number of farmers' markets rose to 5,274 in 2009, up from 1,755 in 1994, and that number continues to climb. The revival of farmers' markets has occurred in Canada and parts of Europe, too. Also, there is a growing interest in community-supported agriculture (CSA), in which individuals pledge support to local farms through subscriptions and receive weekly shares of fruits and vegetables. CSA is thus an alternative economic model of both agriculture and food distribution. It is popular in Canada, the United States, and parts of Europe. In 2010, the United States had about 1,400 CSAs, up from 400 in 2001 and 2 in 1986.[116] The so-called locavore movement extends beyond fruits and veggies, too, as local-minded consumers have rediscovered the benefits of locally brewed beers, locally produced wines, locally raised livestock, and locally produced clothing and fabrics.

The pan-localism of the sustainability movement has dovetailed with other, philosophically like-minded movements since 2000: the bioregionalist movement (which increases economic and cultural cooperation along geographic rather than political lines), voluntary simplicity, vegetarianism and veganism, the slow food movement (eating native, local foods; the opposite of "fast food"), the degrowth movement, and the transition town movement. The latter is a particularly interesting and increasingly well-known instance of localism. The concept of "transition towns" was developed in the British Isles by a permaculturalist named Rob Hopkins. The idea is to create community networks, which work in a bottom–up fashion, that help transition towns and cities toward sustainability. There are now many transition towns throughout England, Sweden, Brazil, and North America. In addition to powering down (using less energy) and building sustainably, transition towns are rooted in the idea of "localization and resilience," which means that they emphasize the local economy, self-sufficiency, and sustainably produced foods.[117]

Hopkins has also been involved in the creation of local currencies, which have become popular in England. These alternative currencies can be purchased with regular money but can be spent only in local establishments that participate in the program, in the hopes of encouraging local consumption. Examples of these new currencies include the Lewes Pound, the Brixton Pound, the Totnes Pound, and the Bristol Pound. England, in many ways, is leading the way in the revival of localism, and parliament has even gotten involved via the Sustainable Communities Act of 2007 and the Localism Act of 2011.

Despite the growth in the local food movement and the renewed interest in local economies, many hurdles remain for sustainable local living. To begin, the market share of local foods is still dwarfed by foods that travel long distances. Direct-to-consumer sales of local foods in the United States accounted for only 0.4% of total agricultural sales in 2007, even though that number was 0.3% in 1997. If one brackets out nonedible agricultural products, direct-to-consumer sales accounted for 0.8% of all agricultural sales in 2007.[118] Second, not all local foods are grown organically or sustainably, which means that local food can still suffer from the same problems as mainstream agriculture. This fact raises ethical dilemmas about whether it's best to buy local conventional foods or organic foods that come from afar. Moreover, many cold-weather areas in North America rely on energy-intensive greenhouses to produce foods during autumn and winter months, and in some cases foreign foods have proven less impactful than local foods. In some ways, warmer-weather locations, with year-round growing seasons, have, at least in theory, a leg up on cold-weather regions when it comes to sustainable local food production, although water consumption is also a problem in many places, such as the US Southwest. Third and finally, there is an ongoing debate about whether eating meat is sustainable, even if that meat is produced locally and organically. In general, animal farming requires extraordinary amounts of land, grain, water, and fossil fuels. It may well be that the only truly sustainable diet is largely or uniquely organic and vegan.[119]

GOVERNMENT PLANNING AND ENVIRONMENTAL POLICYMAKING

It's difficult to summarize governmental support for sustainability since commitment to the cause has varied so widely from one place to the next, from one year to the next, and between the federal, state/provincial, regional, and municipal levels. Certainly, some governments have proven themselves more committed to sustainability than others. Beginning in the 1990s, various governments began to adopt sustainability action plans, which were meant to guide policy in everything from transportation and building to energy consumption, waste, and emissions. The Netherlands led the way with its National Environmental Policy Plan, which has been revised numerous times over the past 20 years. More recent examples include the European Union's 2008 Sustainable Consumption and Production and Sustainable Industrial Policy, which is meant to guide high-level policy throughout Europe, and the city of Vancouver's ambitious "Greenest City: 2020 Action Plan," which aims to transform the Canadian city into the "greenest city in the world."[120] Numerous other regions and cities in the Western world have placed sustainability on the agenda, too.

In fact, governments have done much in recent decades to promote sustainability. The preservation of national parks, natural areas, forests, and marine zones is perhaps the strongest indicator of governmental interest in environmental sustainability. Costa Rica leads the world in percentage of protected territory with 26%. The European Commission made the "precautionary principle" law in 2000, which prevents the implementation of policies that would cause public harm. The clean air and water acts adopted by many federal governments, in addition to new commitments to sustainable forestry, in such countries as Australia, is another sign of growing interest in sustainability. Most UN states are also involved in international environmental treaties, such as the Kyoto Protocol, which aims to reduce GHG emissions below 1990 levels. As we have seen, the UN has been heavily involved in sustainable development since the 1980s. Governments have supported cap-and-trade systems, too, from the United States's sulfur dioxide trading system to the European Union's Emissions Trading Scheme. Countries such as Germany have created market-altering mechanisms, such as eco-taxes and feed-in tariffs, which are meant to promote sustainable production and renewables. In fact, according to Jeremy Rifkin, "More than fifty countries, states, and provinces have 'feed-in tariffs,' which offer producers of renewable energy a premium price above market value for green electricity they sell back to the grid."[121]

On balance, however, there is much more that governments could be doing to support sustainability, and few sustainists are satisfied with governmental support—especially at the national level. The most obvious step would be to generate and actually stick to sustainability action plans. The problem, in many instances, is that most governments are complex jumbles of values and agendas, and thus it is not uncommon for governments to support, for instance, renewable energy, public transit, and sustainable consumption while *at the same time* maintaining outdated pro-growth economic policies that have engendered an unsustainable industrial society. Even the greenest of polities suffer from these contradictions and inconsistencies. The city of Vancouver, for instance, lauds the fact that it has the greenest building code in North America, receives 93% of its electricity from renewables (mainly hydro), and supports green mobility; at the same time, however, the city seems proud of its shocking population growth in recent years (27%), which has increased the city's overall consumption rates and cost of living. Rees, who happens to live and teach in Vancouver, has acknowledged for years that Vancouver has a large ecological footprint and falls well short of its sustainability goals.

Further, international environmental accords are only effective if the big polluters show their commitments to emissions reduction. Currently, the United States and Canada are not actively involved in the Kyoto Protocol. Most countries around the world are still wedded to the fossil fuel paradigm. A look at government energy subsidies reveals that the industrialized world still values fossil fuels above all else. Between 1994 and 2009 (the most recent data), US oil and gas industries received a total of nearly $450 billion in subsidies, compared to the renewables industry, which received a scant $5.93 billion in subsidies over the same 15-year period.[122] The subsidies imbalance needs to be reversed if industrial society hopes to survive and flourish in the long term.

Finally, the commitment to environmental protection at the national and international level from the 1970s to the 1990s, discussed above, masks the many ongoing environmental problems that are, on the whole, poorly managed or even ignored by national governments. Issues such as overfishing, the destruction of old-growth forests, the loss or degradation of soils, the spread of deserts, the growing evidence of endocrine disruption, the malignant sprawl of cities, the loss of biodiversity and species extinction, ecosystem loss or oversimplification, the unsustainable use of water, and the depletion of resources driven by hyper-consumerism are all components of the world's deepening ecological crisis—a crisis that the political sphere has done relatively little to prevent. Ecologists such as C. S. Holling and his associates in the Resilience Alliance have devised new forms of environmental management that reconceptualize how states can approach stewardship, restore natural capital, and safeguard essential ecosystem services, but these ideas have only recently begun to gain attention and credibility.[123] When it comes to addressing the sustainability movement's biggest concerns, national governments and the international community still have a long way to go.

As we've seen in this chapter, sustainability is a multifaceted movement that has become prevalent in a range of institutions, practices, and scholarly fields. Since the 1990s it has decentralized and expanded beyond its institutional base within the United Nations. As a set of principles, sustainability has provided guidance and vision for such sectors as transportation, urban planning, energy, food production, and the built environment. It has also served as the foundation for new kinds of data and new tools that have transformed sustainability from a nebulous abstraction to a set of identifiable processes and outcomes. Sustainability is now a measurable "state." It is a growing academic field (or set of fields), a form of expertise, and a viable career path. As the movement has grown and gained legitimacy, it has provided the industrialized world with a viable alternative to the narrative of modernity that has shaped the West since the eighteenth century. Sustainists offer a blueprint for a sustainable society and a green economy that functions prosperously within ecological limits and values life satisfaction over growth and consumption.

Chapter 7

The Future:
10 Challenges for Sustainability

Growing concerns about climate-change pollutants, the widening gap between the rich and the poor, resource shortages, and the world's gamut of ecological problems have placed new pressures on sustainists. Creating a sustainable society that thrives within its biophysical limits is no longer seen as a distant and utopian objective; it's now an urgent matter that, if neglected or mismanaged, will bring devastating consequences for the planet and the human economy that lives off of it. The increased political attention, institutional support, and financial commitment to the cause of sustainability means heightened expectations for immediate, tangible results. The public doesn't want idle chatter; it wants workable solutions to very real problems. Can sustainists seize the moment and lead the transition to the sustainable future?

The quest to create a sustainable society faces a host of obstacles, and many pressing questions remain unanswered: How can the entrenched political and corporate interests that perpetuate unsustainability be overcome? How can society willingly transform itself? Where will the money and political will come from to coordinate the transition? Will this sustainable society be "industrialized" or "post-industrial," "globalized" or "localized"?[1] Will the changes be top–down, bottom–up, or both? By charting the growth and development of sustainability since 1700, this book has not meant to imply that ecotopia is an inevitable end point. Even optimists concede that it's quite possible that the task is too tall, that industrial society could drive itself straight into the ground, that collapse is a real threat, and that the Industrial Revolution was the first phase of humanity's protracted extinction event. If sustainability does succeed in undoing the many harms that have caused our ecological predicament, it will only do so with the broad support of the public and through a cooperative effort to adapt and transform. At the risk of bombast, it will have to change the course of human history, and that's no easy task.

This book ends with a discussion of 10 challenges faced by the sustainability movement. There are certainly others to mention, but these 10 are the most commonly cited in the literature on sustainability and, sustainists argue, the most pressing.

CREATE A SHARED VISION FOR THE FUTURE— AND STICK TO IT

Creating solutions to complex problems requires broad cooperation and even consensus, but ever since sustainability went mainstream and expanded its scope, it has sheltered a messy blend of interests and viewpoints. There were always disagreements among sustainists on the question of economic growth. But sustainability has become an even more fraught and contradictory catch-all movement in recent years, especially within the world of administration and policymaking. My own experiences working on university sustainability committees anecdotally confirm the point. These committees often bring together Chicken Little survivalists ("the sky is falling!"), business-as-usual types who simply want to maintain the status quo and call it "sustainability," and a range of pragmatic middle-of-the-road strategists, each of whom pushes some particular agenda. This range of viewpoints likely mirrors the divisions of the wider society. Multiple perspectives is certainly a positive thing to have, but the first and most important challenge of the sustainability movement is to get people on the same page.

Transforming society from industrialized unsustainability to a social and economic system that is sustainable requires a consensus-building and viable blueprint for the future. In the words of Gwendolyn Hallsmith and Bernard Lietaer, "A long-term vision guided by people's values—what they care deeply about—is a prerequisite for mobilizing collective action."[2] But who develops the vision? What happens when there are multiple, competing visions, and who decides which one to choose? Since the 1970s, a host of future-oriented thinkers, schooled in ecology, community planning, and/or ecological economics have crafted visions of a sustainable society. We've met many of these thinkers in the pages of this book: David Holmgren, Herman E. Daly, Rob Hopkins, Jeremy Rifkin, John Ehrenfeld, Paul Hawken, Amory Lovins, Hunter L. Lovins, Richard Heinberg, Lester Brown, Eben Fodor. Ian Lowe and David R. Boyd have each published detailed outlines as have many other sustainists.[3]

Broadly, the vision of these thinkers is of a low-carbon, powered-down, and ecologically sound economy buttressed by decentralized democratic institutions.[4] The challenge is not only to harmonize different visions for a green future but also to convey effectively the message of sustainability to the broader society. There is a need, moreover, to have not only a vision but the "grunts" to make it happen.[5] The sustainability movement will succeed only if the vast majority of ordinary people come to see the idea of living sustainably as important and relevant to their lives. This narrative needs to be honed, broadcast, and adopted. As John Ehrenfeld argues in *Sustainability by Design*, "Sustainability can emerge only when modern humans adopt a new story that will change their behavior such that flourishing rather than unsustainability shows up in action."[6]

MOVE PAST NEOCLASSICAL ECONOMICS, DEREGULATION, AND THE GROWTH OBSESSION

When the G8 met at Camp David in 2012, virtually the only thing that the world's leaders could agree upon was the need for more economic growth—this, despite unprecedented levels of throughput, shocking degradation to global ecosystems, rising populations, high unemployment, and record levels of climate-altering pollutants. The world economy is the biggest it's ever been in history. In 2012, the world's gross domestic product (GDP) stood at over $71 trillion. Not only has constant growth not solved the world's social and environmental problems, but sustainists argue that growth exacerbates social inequalities and leads to despoliation, pollution, and overconsumption. And yet, for most economists and the political class, growth remains an unscrutinized article of faith. It's the dogma of the "market fundamentalists," and since the idea is so fundamental to neoclassical economics, it tends to be defended with great vigor.[7]

The other article of faith in modern economics, more shaken of late, is the neoliberal's belief that the deregulation of markets creates stability and economic benefits for everyone. As the BBC's Paul Mason has shown, the great recession of 2007–2009 was caused by the deregulation of financial markets that culminated in the United States in 1999 under the Clinton administration.[8] In that year, the Glass–Steagall law of 1933, which had separated investment banks from savings banks and had helped the United States recover from the Great Depression, was quietly repealed. In its place came the Gramm–Leach–Bliley Act, which overturned Glass–Steagall, followed shortly thereafter by the Commodity Futures Modernization Act, which removed all regulation from the volatile derivatives market. The result, according to Mason, was "deregulated investment banking, expanded subprime mortgage lending to the poor and ethnic minorities, a planet-sized derivatives market, and the fusion of banking and insurance."[9] The whole deck of cards collapsed in 2007–2008, crushed under the weight of massive debt and facilitated by a deregulated economy and an offshore "shadow banking system." There is now an abundance of evidence to show that regulation brings stability to markets, and 2007–2008 is merely the latest—but probably the largest—example of why deregulation hurts the poor and the wider economy. As Paul Hawken noted pithily after the meltdown, "A free market, so lovely in theory, is no more feasible in practice than a society without laws."[10] Regulation of markets has never looked like a better idea, and the economies that possessed tightly regulated banking systems in 2007–2008, including Canada, largely avoided the meltdown.[11]

Growth and deregulation have become central to modern industrial capitalism, and thus to jettison these dogmas requires a trip back to the drawing board. Sustainability economists have responded, as we have seen, by developing alternative, non-neoliberal forms of capitalism that reject the need for more material growth, more debt, and more economic corpulence and that define prosperity in terms of well-being, flourishing, life satisfaction, and ecological equilibrium.[12] Building a green and intelligently regulated economy is perhaps *the* challenge of the present day. For over 40 years, Herman Daly has been the most vocal critic of the hypocrisies and oxymorons of "managed growth, balanced growth, smart growth, green growth" and all the other "temporizing evasions" of neoclassical economics.[13] His concept of steady-state economics forms the foundation of a largely untested model of macroeconomics that, in theory, would establish an industrialized society that is stable and prosperous and operates within its biophysical limits.[14]

As an antidote to the growth obsession, "degrowth" has become a commonplace word and concept within the world of sustainability.[15] Degrowth is an anti-consumerist movement that seeks to reverse the trend of economic expansion by deliberately powering down society, rediscovering local economies, and promoting non-materialistic lifestyles. Heinberg's *Powerdown*, Lewis and Conaty's *The Resilience Imperative*, Fodor's *Better Not Bigger*, Ted Trainer's *The Transition*, and Hopkins's *Transition Companion* are all books that lay out practical steps for transitioning away from energy-intensive industrialism to more sustainable and localized forms of life.[16] Ozzie Zehner's *Green Illusions* and Trainer's *Renewable Energy Cannot Sustain a Consumer Society* remind us that growth in material consumption—which is the desired outcome of neoclassical economic policies—is unsustainable and that renewable energy makes sense for modern society only if energy demands and consumption levels are first greatly reduced.[17] Finally, as we saw throughout this book, the challenge for economists is to move past the unhelpful tools of neoclassical economics, from the GDP to cost–benefit analysis, and mobilize methods that measure and promote equality, well-being, resource efficiency, and ecological principles. A new macroeconomics for a sustainable economy needs to emerge.

FACE SHORTAGES, BECOME RESILIENT

Human society needs to adapt quickly to the changing realities of our world, while still doing its best to mitigate ecological disasters.[18] The climate is already in the process of warming, which, we now know, has a devastating impact on agricultural productivity, disease, weather patterns, and coastal cities. Fresh water resources are growing scarce, topsoil is becoming depleted, and groundwaters are badly polluted. Peak oil has already occurred. We have overspent from the carbon bank buried beneath our feet. There is a growing scarcity of metals in the world's mines, too; copper reserves, for instance, may be depleted as early as the 2030s, and platinum, hafnium, indium, gallium, and zinc are running low.[19] The population of homo sapiens is now above 7 billion—the highest it has ever been in our 200,000 years on the planet, making it ever more difficult to feed the human race. According to those trained in systems ecology, humanity is facing a massive crisis, and one that will require quicker-than-normal rates of species evolution.

The ecological concept of "resilience" refers to the capacity of ecosystems (or species) to cope with change. The extrapolated lessons that come from this science seem increasingly germane for human beings. What ecologists have found is that species that are best able to "absorb disturbance," adapt, and transform are those most likely to evolve successfully.[20] Social scientists and ecological economists have taken the concept of resilience and applied it directly to humanity's predicament. According to Lewis and Conaty, the seven principles of resilience are diversity, modularity, social capital, innovation, overlap, tight feedback loops, and ecosystem services. Understanding and valuing these principles gives humanity the best chance of surviving ecological crisis.[21]

On a more practical and immediate level, resilience means coping with the present and planning for the future. It means ensuring genetic diversity and hardiness in our agricultural systems. It means radical forms of recycling that eliminate waste. It means relying on sustainably managed forests and fisheries. It means making do with less stuff. It means building structures and cities that can cope with storms, rising sea levels, and extreme heat and cold. It means using coal, oil, and gas as transitional fuels en route to an economy based on renewables and low rates of consumption. Human society can no longer function with the assumption that levels of consumption—of fossil fuels, of metals, of food, of plastic, and so on—will continue to rise as they have for the past several decades. We are moving quickly toward the limits of growth—indeed, we are already living in global overshoot, according to Ecological Footprint Analysis. Coping with food and other resource shortages and volatile weather systems will test the human capacity for creativity and adaptation, especially if, as many scientists project, the average planetary temperature rises by 6° to 8° F over the course of the twenty-first century.[22]

HARMONIZE THE NEEDS OF RICH AND POOR, THE DEVELOPED AND THE DEVELOPING

Sustainability cannot become a movement that is only affordable for and attractive to the wealthy people of developed countries. Van Jones's call for "more eco-populism" and "less eco-elitism" gets to the heart of the matter.[23] The concept of living green will be undermined if it is associated solely with wealthy, high-profile celebrities whose energy-intensive, high-consumption lifestyles seem at odds with their message of ecological wisdom. As it is, buying local organic food and installing private solar panels is difficult for the average person to afford. Moreover, being green is often seen as something distant and unrelated to the lives of the working and urban poor. Jones's argument is that sustainability can succeed only if it is accessible and if its importance is made obvious for all social classes. It must have broad appeal, especially since ecological problems affect everyone.

In terms of international relations, there is also an urgent need to harmonize approaches to sustainability in and between developed and developing countries. Sustainability should not be seen as something that a wealthy country achieves after having gone through a messy process of industrialization. In fact, it is in the world's poorest countries where developing toward sustainability is most pressing. As Al Gore writes in *The Future*:

> The consequences that are already occurring, let alone those that are already built into the climate system, are particularly devastating to low-income developing countries. Infrastructure repair budgets have already skyrocketed in countries where roads, bridges, and utility systems have been severely damaged by extreme downpours and resulting floods and mud slides. Others have been devastated by the climate-related droughts.[24]

Jared Diamond, in *Collapse*, reacts with great hostility to wealthy Westerners who assume that "environmental concerns are a luxury affordable just by affluent First World yuppies, who have no business telling desperate Third World citizens what they should be doing." Diamond continues:

> In all my experiences of Indonesia, Papua New Guinea, East Africa, Peru, and other Third World countries with growing environmental problems and populations, I have been impressed that their people know very well how they are being harmed by population growth, deforestation, overfishing, and other problems. They know it because they immediately pay the penalty,

in forms such as loss of free timber for their houses, massive soil erosion, and (the tragic complaint that I hear incessantly) their inability to afford clothes, books, and school fees for their children.

Sustainability is not an end point for the West but rather a process that could and should be shared globally. The developing world does not need to make the same mistakes or follow the same development path in fast forward as the Western world did with its Industrial Revolution.

Finally, Jeffrey Sachs, along with climate and environmental scientists, has shown that the world's ecological crisis affects the entire globe. There is no escape and no hiding, even if the coming changes will not affect all parts of the world in the same ways or at the same time. Yet global climate change, loss of biodiversity, impairment of ecosystems, and a growing population are threats to *all* human societies. We are all downstream from one another. Sachs uses this argument to defend the financial and logistical aid that wealthier countries should give to the developing world. The logic is that helping others is not only ethical and altruistically good but also creates global stability that is ultimately beneficial for the industrialized world, too.[25]

The interconnectedness of global trade, energy, food, and financial networks means that systems disruptions—or rather, unsustainability—eventually impacts everyone. We learned this lesson most recently when the economic meltdown of 2007–2008 went global, and then again in 2008 when commodities prices spiked and drove the world's poorest people to the brink of famine.[26] Also, various climate-related and ecological problems—especially droughts and floods—have already given rise to a growing number of "climate refugees" who are moving away from parts of Africa, the Middle East, Asia, and elsewhere in search of safety and stability.[27]

RETHINK ENVIRONMENTAL MANAGEMENT, SAFEGUARD ECOSYSTEM SERVICES, AND RESTORE NATURAL CAPITAL

Many sustainists differentiate between "weak" and "strong" sustainability. Weak sustainability means that humanity moves away from its destructive habits and manages to stabilize ecosystems, the climate, and the human population. Strong sustainability goes a step further and demands that humans actively *repair* much of the environmental damage that they have done (to the extent that this is possible). Strong sustainability is particularly influenced by the theory of ecological resilience, mentioned above. For resilience experts, governmental and industrial approaches to environmental management need to change in at least three ways, with the ultimate objective being the sustainable use of natural resources.

First, the outdated notion of maximum sustained yield, which has been employed in different ways from forestry to fisheries, needs to be scrapped altogether, given the spotty track record of this resource management method in maintaining sustainable rates of harvest. The problem is that Western science has tended to assume that ecosystems are static, simplistic, and easily managed. It also assumes that humans can quickly "fix" any error in the system—such as overharvesting. In fact, natural systems have often been "engineered" in ways that simplify them and make them more vulnerable to collapse. Rees explains: "When humans maximize the harvest of a particular species, for example, we inadvertently alter the species' relationship to multiple other species (e.g., predators and prey) in the ecosystem, setting off a cascade of feedback responses that can fundamentally erode the system's integrity."[28] In short, when an ecosystem is engineered to produce a maximum yield, it loses resilience and collapses, just as the Atlantic Northwest Cod fishery did in the late 1980s and early 1990s. Holling and other ecologists argue that resource managers need to understand the "adaptive cycles"—the flux—that governs the growth, maturity, collapse, and reorganization of ecosystems.[29] For Rees, the solution is to shift resource management from maximization to "sufficiency" and to work with rather than against natural processes.[30]

The second way in which environmental management needs to change is by concentrating on the safeguarding of crucial ecosystem services. Ecosystem services are all those natural processes that allow for a healthy and functional environment and that, by extension, ensure the stability of our societies and economies. Examples include the climate's ability to regulate the temperature of the planet, the pollination of crops by bees, the purification of water and air, the decomposition and detoxification of waste, the dispersal of seeds, and the ability of nutrients to disperse and circulate through an ecosystem. Focusing on ecosystem services can give environmental stewards a clearer idea about *what* needs protecting.[31]

Third and finally, sustainists urge governments and resource managers to go beyond mere protection and sustainable consumption to the *restoration* of natural capital. This means undertaking active efforts to clean up polluted waterways, reforest denuded areas (with species biodiversity in mind), rebuild the humus content (fertile, organic matter) in depleted soils, and help endangered species to come back from the edge.

CLIMATE CHANGE IS A GIGANTIC PROBLEM . . .

There is an overwhelming scientific consensus that (i) the global surface temperature is rising steadily and (ii) that humankind's activities, especially the burning of fossil fuels and the clearing of forests (and other land use practices), are primarily responsible for this steady increase in temperatures. The planet has warmed by 1.08°

F (0.6° C) over the past few decades and 1.44° F (0.8° C) since the early twentieth century. According to James Hansen, the top climate scientist at NASA from 1981 to 2013, surface temperatures are now rising by about 0.36° F (0.2° C) per decade. Surface temperatures in 2100 could be, on average, 8.1° F (4.5° C) higher than they were in 2000. The problem is that there is more carbon dioxide (CO_2), methane, and nitrous oxide in the atmosphere than there once was. Carbon dioxide is, at least for now, the biggest problem for the climate, since all the "new" (once buried) carbon that has been released from burned fossil fuels (since the eighteenth century) has radically altered the carbon cycle. The atmosphere in pre-industrial times contained 280 parts per million of CO_2, and thus there was a more or less stable amount of carbon in the atmosphere. But by 2013, the parts per million of CO_2 had surpassed 400. All of these new carbon molecules act like a blanket that traps heat in the atmosphere. The bonds that hold the molecules together absorb and radiate energy at infrared wavelengths, which prevents energy from flowing from the Earth's surface up to space. The year 2010 was the hottest in recorded history, and 2012 broke many more heat records.[32]

Moreover, the climate suffers from many "positive feedback loops" that accelerate the rate of temperature change and intensify its effects. In other words, the problem simply compounds over time. First, the world's oceans are becoming more acidic and warmer and are thus less able to absorb CO_2. The oceans are an important "sink," along with forests, that must be able to absorb and process the excess carbon in the atmosphere. Second, the warming of tundra and permafrost is contributing to the release of natural stores of methane gas, which is many times more harmful for the climate than CO_2. Third, a process called "ice-albedo feedback" is occurring as the polar ice caps melt. Those regions become less reflective and thus absorb more solar radiation, which speeds up the process of warming. Climate change can also intensify the spread of deserts and kill off forests, which, aside from being hugely important resources, are needed as carbon sinks to fight the process of warming.

The effects of climate change are already being felt around the world: severe droughts, such as the one that hit North America in 2012–2013; relentless storms, such as 2012's Hurricane Sandy (United States) and 2013's Typhoon Haiyan (Philippines); and the early arrival of spring in parts of the Northern Hemisphere, which is a problem that has a largely negative impact on ecosystems. But the worst is yet to come. Scientists argue that the effects of climate change, especially if it is allowed to spiral out of control throughout this century, will be severe and far reaching: sea levels will rise in coastal areas in many parts of the world, causing flooding in low-lying areas; the ocean will become more acidified, contributing to the decline of coral reefs and many fish species that feed billions globally; many terrestrial species will become extinct or exceedingly rare from habitat loss, overhunting, and other issues; the world's ecosystems will become destabilized and further degrade; the frequency and intensity of storms will increase; droughts and weather inconsistencies will reduce agricultural outputs in corn and other staples (affecting food availability and prices); and infectious disease, weeds, and insects will thrive. In short, it will be very, very bad. Dealing with the climate change that has already occurred and doing what we can to reverse the process will be an immense cultural, economic, and political challenge, and one that will be regarded by those of the future as a major crossroads for humanity.

... BUT IT'S NOT THE ONLY ONE

As if climate change weren't a big enough problem, there are also many other ecological issues that threaten the creation of a sustainable society. An article from 2007 by Eileen Crist makes the case that the public and policymakers should not focus solely on climate change.

> While the dangers of climate change are real, I argue that there are even greater dangers in representing it as the most urgent problem we face. Framing climate change in such a manner deserves to be challenged for two reasons: it encourages the restriction of proposed solutions to the technical realm, by powerfully insinuating that the needed approaches are those that directly address the problem; and it detracts attention from the planet's ecological predicament as a whole, by virtue of claiming the limelight for the one issue that trumps all others.[33]

For Crist, climate change has merely exacerbated preexisting ecological problems, and even if the climate challenge were magically solved, the world would still face a number of significant crises rooted in the destructive patterns of a growth-based, consumer-capitalist society. "The problem is a sprawling civilization that is destroying the biosphere, and will continue to do so even after it (somehow or other) deals with a major glitch in the machine—the consequences of accumulating greenhouse gases."[34] Issues such as "species extinction, overfishing, destruction of old growth, topsoil loss, desertification, endocrine disruption, endless development, [and] biodiversity crisis" are all problems that exist independently of the climate crisis, even if the latter tends to make them even worse.[35]

What's more, as we have seen in this book, sustainability also addresses social and economic issues, from critiques of growth-based economics to the fight for social justice. The sustainability movement should not pigeonhole itself as a one-item cause that deals uniquely with climate change. To this end, many sociologists, environmentalists, ecologists, and economists have begun to shift the central focus of the discourse toward the root causes of ecological problems, the most significant of which appears to be overconsumption. As Ted Trainer and many others have argued, consumerism presents an almost insurmountable hurdle for sustainability. The throughput, waste, and energy consumption that comes from the globalized trade in stuff makes it nearly impossible to establish a zero-waste, ecologically balanced, and equitable society that runs on renewables.[36] As a result, the message of many sustainists is to slow down, power down, simplify, and, above all, consume less. For Holmgren, we need to adapt quickly to an "energy descent society" based on low consumption and a high degree of self-sufficiency.[37]

FIGHT GREENWASHING AND THE DENIAL INDUSTRY

One of the biggest fears about the expansion of the sustainability movement is that it will enable "greenwashing"—the use and abuse of sustainability language or imagery to mask conventional, destructive practices. Greenwashing is a way of exaggerating or fabricating the environmental benefits of a product, practice, or service—and indeed many corporations have greenwashed their own products through buzzwords and misinformation in an effort to tap in to the growing market in "green products." This is problematic for two reasons. First, it makes a mockery of the concept of sustainability, and, second, it is merely another way for marketers to mislead consumers and exploit benign intentions. One report on greenwashing from 2010 lists seven "sins" of greenwashing—hidden trade-offs, an absence of proof, vagueness, irrelevance, the lesser of two evils, fibbing, and the worshipping of false labels—and found that a whopping 95% of "greener" products commit one or more of the seven sins.[38] Even though, per the report, greenwashing has gone down over time, as green products are more vigorously policed, there are still many cleaning products, for instance, that claim to be "green," "natural," or even "organic" but that use many of the same chemicals as non-green products. The fight against greenwashing, which has become more prevalent in recent years, is a way of empowering the public and ensuring honest forms of marketing.[39]

Greenwashing is a way of exploiting sustainability. A more radical approach has been to undermine its credibility altogether, along with the ecology, systems theory, and climate change science upon which it draws. This tactic became so prevalent that in 2006 the journalist George Monbiot popularized the term "the denial industry" to label the systematic and purposeful attempt on the part of corporations and ultra-conservative organizations to sow doubt into the dialogue on climate change. "For years, a network of fake citizens' groups and bogus scientific bodies has been claiming that [the] science of global warming is inconclusive."[40] ExxonMobil and other oil companies were eventually exposed as the main backers of the denial industry. Monbiot continues:

ExxonMobil is the world's most profitable corporation [in 2006]. Its sales now amount to more than $1bn a day. It makes most of this money from oil, and has more to lose than any other company from efforts to tackle climate change. To safeguard its profits, ExxonMobil needs to sow doubt about whether serious action needs to be taken on climate change. But there are difficulties: it must confront a scientific consensus as strong as that which maintains that smoking causes lung cancer or that HIV causes Aids.

In the late 1980s, NASA and the Intergovernmental Panel on Climate Change agreed on the human causes of climate change and urged the US Congress and other governments to take action to curb emissions. In the words of Ozzie Zehner:

> Feeling threatened, several oil companies and other large corporations joined forces to fund advertising campaigns, foundations, and organizations such as the American Enterprise Institute, the Global Climate Coalition, and the George Marshall Institute, in order to attack the credibility of scientists studying climate change and to frame climate change as a scientific "dispute" rather than a consensus. These organizations hired many of the same public relations and legal consultants who had earlier ridiculed doctors for warning about the risks of cigarette smoke.[41]

In the end, the "merchants of doubt" proved to be a huge success.[42] Their campaigns "swayed public opinion, greatly influenced media coverage, and delayed policy to mitigate the effects of climate change." In the first decade of the twenty-first century, most Americans accepted the reality of climate change but also believed that scientists were still making up their minds about its causes and other aspects of the science, when in fact the basic scientific knowledge had been established since the late 1980s.[43]

Greenwashing and the denial industry are very significant challenges for the sustainability movement. They serve as potent reminders that those who profit from unsustainable industrial practices will use underhanded tactics to further their business interests—even if those interests stand in direct contrast to the long-term interests of the global community. The denial industry, in particular, is driven largely by the malice of a self-interested fossil fuel industry seeking to curtail regulation on energy extraction standards, emissions levels, and consumption patterns. Meanwhile, the parts per million of CO_2 continue to climb. The confusion and doubt created by the denial industry serve the interests of humanity not at all and are merely a crude and selfish attempt by certain corporations to continue to profit from our ecological crisis.

GALVANIZE PUBLIC SUPPORT AND POLITICAL ACTION WITHOUT GETTING POLITICIZED

A second fear is that sustainability will become politicized and associated with particular political parties, thereby undermining its ability to float above the political fray. It is somewhat surprising that sustainability has mostly avoided unhelpful politicization, even as it's become a permanent feature of United Nation's policymaking, and even though political parties sometimes explicitly support sustainability. To my knowledge, though, there is no "Sustainability Party" anywhere on the planet, and as a discourse, it is still relatively safe territory for the right, left, and center, although the right in countries such as the United States and Canada tend to invoke the concept less than their political rivals. However, in the United Kingdom, the Conservatives, the Labour Party, and the Liberal Democrats all invoke the language and concepts of sustainability without one party being seen as having exclusive possession of the discourse.

If the sustainability movement hopes to achieve real change, then it must galvanize public support without becoming associated with an established political ideology or party. Sustainists have been largely successful at projecting political neutrality by conveying the idea that living sustainably is good for *everyone*. The broad interests of humanity and the well-being of the planet transcend the narrow and self-interested ethos of political parties. Thus the challenge is to have a politically active movement without becoming *politicized*.

FINANCE THE REVOLUTION

The transition to a clean, low-carbon economy will require heavy financial investments. Someone needs to pay for all those bike lanes, high-speed trains, renewable energy projects, and recycling plants. The problem is that the reigning venture capital approach demands large and immediate returns on investment. This model might be well suited to certain domains of the economy, but it makes little sense for the long-term prospects of the green economy. Much of the "clean tech" that will facilitate the transition to a sustainable society, such as third-generation micro-algae biofuels, new forms of waste-free production, and hydrogen storage, require years of research and development, involve large capital commitments, and yield little or no immediate returns on investment. In fact, some may prove to be dead ends altogether. As a result, investors often balk at financing clean tech, and, in fact, worldwide clean tech investments declined drastically in 2012.[44] The clean energy sector seems to be in the process of bouncing back, and there was still five times more invested in clean tech in 2012 than there was in 2004, but it's clear that some potential investors are put off by the risks and the lack of immediate returns.[45]

New models of investment need to emerge if the green economy is destined to succeed.[46] The three most common suggestions for financing sustainability are these. First, locate and develop reliable forms of capital investment. For Jeremy Rifkin, this means further developing microfinance, selling the benefits of clean tech industries to traditional investing agencies, and securing as much governmental financial support as possible. If Western governments bankrolled the Industrial Revolution, they can also finance the transition to a new, more sustainable economic mode.[47] But the money needs to come from somewhere.

Second, give it time. Paul Mason not only argues for re-regulating the financial world in his analysis of the 2007–2008 meltdown but also offers a vision of a "low-profit, utility-style commercial banking" system, in which quarterly profits and short-term gain are set aside in favor of long-term, low-yield investments that benefit society and the environment.[48] In short, the new investing model cannot be driven solely by immediate profitability but must take into account non-market considerations, from social justice to resource conservation and pollution levels.

Third, shift investments and subsidies from unsustainable fossil fuels to sustainable forms of energy. The "divestment" campaign, in which universities and other organizations have gotten rid of their fossil fuel investments, is one example of shifting investments.[49] But the bigger battle is now being fought over governmental subsidies for fossil fuels—all those direct funds, tax breaks, materials, and infrastructure projects that keep the cost of fossil fuels artificially low. It is famously difficult to determine the amount of money that the fossil fuel industry saves due to governmental support, but a 2010 report by the Global Subsidies Initiative estimates that the fossil fuel industry receives at least $100 billion per year worldwide, and likely much more if ancillary forms of support are factored in to the equation.[50] To better finance the sustainability revolution, taxpayer money must shift away from the things we don't want—pollution, nonrenewable fuels—and toward the things that we do—a clean and sustainable economy that is beneficial for all.

Clearly, the sustainability movement faces tough hurdles, but sustainists remain optimistic. Consciousness about the pitfalls of fossil fuels, endless growth, social inequalities, and industrialization has never been greater, and there is, increasingly, a broad consensus that the global community needs to adopt a new narrative, one that allows our species to live safely, comfortably, and in harmony with the natural environment. The fairytale of industrial-growth-as-progress thrust upon us since the eighteenth century is basically dead. Indeed, the violent storms of 2012 and 2013, the troubling temperature records since 2010, the ruinous economic meltdown of 2007–2008, and the sobering natural resource shortages of the early twenty-first century have made the message of sustainists difficult to ignore and the lingering dogmas of growthists[51] difficult to stomach.

The purpose of this book has been to demonstrate that the fundamental idea of creating a sustainable society has a long genealogy that stretches back at least to 1700. It was also written to connect past, present, and future. By looking backward, we are better able to understand the present and the future—a future that has become an object of dire concern for the citizens of the twenty-first century. Indeed, looking back at the slow development of the concept and then the movement of sustainability is beneficial for at least two reasons. First, it presents us with a useful critique of capitalist industrialization and its impact on global societies and ecosystems. Second and more constructively, it offers practices, technologies, and bodies of knowledge that can move society and the economy in a different and more viable direction.

Finally, the sustainable society envisioned by so many is, in many ways, a modernized revival of past wisdom, much of which was drowned out in the rush for industrial growth. Sustainability borrows from the localism and seasonal diets of pre-industrial societies. It adopts conceptions of renewable resource harvesting that trace back to feudal Japan and early eighteenth-century Saxony. It makes use of mid-nineteenth-century views of a steady-state economy. It retains the desire for social justice and equality articulated by the myriad critics of the Industrial Revolution. We can never truly return to some pre-industrial ecological utopia. But we can pay close attention to the pioneers who shaped the idea of living in peaceful perpetuity upon the Earth. We can listen to the prescient critics who warned of the effects of reckless industrialism. We can observe the path *not* taken and thereby find new ways to rediscover it.

NOTES

Introduction

1. Chris Turner used this expression in a presentation at the University of Alberta in 2010. Also see Turner, Chris. *The Geography of Hope: A Tour of the World We Need*. Toronto: Vintage Canada, 2007.
2. McKibben, Bill. "Buzzless Buzzword." *New York Times*, 10 April 1996. See also Silverman, Howard. "Sustainability: The S-Word." *People and Place: Perspectives*, 15 April 2009. Online: http://www.peopleandplace.net/perspectives/2009/4/15/sustainability_the_s-word.
3. The earliest I've found include Stivers, Robert L. *The Sustainable Society: Ethics and Economic Growth*. Westminster: John Knox Press, 1976; Pirages, Dennis Clark, ed. *The Sustainable Society: Implications for Limited Growth*. New York: Praeger Publishers, 1977.
4. Sachs, Jeffrey D. *Common Wealth: Economics for a Crowded Planet*. New York: Penguin, 2008.
5. Intergovernmental Panel on Climate Change. *The Synthesis Report of the Fifth Assessment Report*. Geneva: IPCC, 2014. It strengthens the language and findings of the 2007 report, *Climate Change 2007: Synthesis Report*. For instance, the 2014 report found that there is a 95% to 100% chance that human actions are the primary cause of the warming of the past few decades, whereas the 2007 put the figure at 90% to 100%.
6. See Hansen, James, Makiko Sato, Reto Ruedy, Ken Lo, David W. Lea, and Martin Medina-Elizade. "Global Temperature Change." *Proceedings of the National Academy of Sciences* 103.39 (July 2006): 14288–14293.

7. Sachs, *Common Wealth*, 139. See also the Millennium Ecosystem Assessment. *Ecosystems and Human Well-Being: Biodiversity Synthesis*. Washington, DC: World Resources Institute, 2005.

8. See, for instance: Mason, Paul. *Meltdown: The End of the Age of Greed*. New York: Verso, 2010.

9. Rifkin, Jeremy. *The Third Industrial Revolution: How Lateral Power Is Transforming Energy, the Economy, and the World*. New York: Palgrave Macmillan, 2011.

10. The definition of social justice is still debated, but it's based on the concept of human rights as defined in the UN's 1948 Universal Declaration of Human Rights. It also implies the ability to flourish, the right to live in a healthy environment, access to basic life necessities, and freedom from oppression.

11. I take inspiration from Hans Ulrich in this quest. He is one of the only contemporary writers who is working on the history of sustainability. See Grober, Ulrich. *Sustainability: A Cultural History*. Trans. Ray Cunningham. Totnes, UK: Green Books, 2012.

12. See, for instance, Edwards, Andres R. *The Sustainability Revolution: Portrait of a Paradigm Shift*. Gabriola, BC: New Society Publishers, 2005.

13. Greenwashing is a form of spin, in which a business or an organization gives the false impression that its products and/or services are environmentally friendly. It's a way of exploiting the public's desire to change its consumption habits. Many firms want to seem green, but in fact are not and therefore come across as liars or hypocrites. Greenwashing makes it difficult for voters, citizens, and consumers to know whom to trust.

14. In French, "sustainability" is ordinarily translated as "la durabilité" (and "sustainable development" is "le développement durable"). In Spanish, it's "sostenibilidad, " and in Italian, "sostenibilità."

15. Daly, Herman E. "Introduction." In *Toward a Steady-State Economy*. Ed. Herman E. Daly. San Francisco: Freeman, 1973, 23. Also see Victor, Peter. *Managing Without Growth: Slower by Design, Not Disaster*. Cheltenham, UK: Edward Elgar Publishing, 2008, 22.

16. For an excellent overview of the principles of sustainability, see Dresner, Simon. *The Principles of Sustainability*. Earthscan, 2008.

17. The Natural Step. "The Four Systems Conditions." www.thenaturalstep.org

18. Heinberg, Richard. "What Is Sustainability?" *The Post-Carbon Reader: Managing the 21st Century's Sustainability Crises*. Eds. Richard Heinberg and Daniel Lerch. Healdsburg, CA: Watershed Media, 2010, 13–24.

19. However, Heinberg does draw on the following work: Jessie Dillard, Veronica Dujon, and Mary C. King, eds. *Understanding the Social Dimensions of Sustainability*. New York: Routledge, 2008.

20. Bartlett, Albert A. "Forgotten Fundamentals of the Energy Crisis." *American Journal of Physics* 46 (September 1978): 876–888; Bartlett, "Reflections on Sustainability, Population Growth, and the Environment—Revisited." *Renewable Resources Journal* 15.4 (Winter 1997–1998): 6–23.

21. Foucault, Michel. *L'archéologie du savoir.* Gallimard, 1969. The book is usually translated as "Archaeology of knowledge" and appears in many editions.

22. Dryzek, John S. *The Politics of the Earth: Environmental Discourses.* 2d ed. Oxford: Oxford University Press, 2005.

23. Worster, Donald. *Nature's Economy: A History of Ecological Ideas.* 2d ed. Cambridge, UK: Cambridge University Press, 1994.

24. The four principles of social sustainability are human well-being, equality, democratic government, and a democratic society (including the paramount concept of social justice). Social sustainability promotes both environmental and economic sustainability. See Magis, Kristen, and Craig Shinn, "Emergent Principles of Social Sustainability." In Dillard, Dujon, and King, *Understanding the Social Dimensions of Sustainability,* 15–44.

25. See Harris, Jonathan M. and Neva R. Goodwin. "Volume Introduction." *A Survey of Sustainable Development: Social and Economic Dimensions.* Eds. Jonathan M. Harris, Timothy A. Wise, Kevin P. Gallagher, and Neva R. Goodwin. Washington, DC.: Island Press, 2001, xxix.

26. Cited on page 204 of Meadows, Donella H., Dennis L. Meadows, and Jørgen Randers. *The Limits to Growth: The 30-Year Update.* White River Junction, VT.: Chelsea Green Publishing, 2004.

27. Diamond, Jared. *Collapse: How Societies Choose to Fail or Succeed.* New York: Viking Press, 2005.

28. Mishan, E. J. *The Cost of Economic Growth.* London: Staples, 1967; Mishan, *Economic Growth Debate: An Assessment.* London: Allen & Unwin, 1977; Meadows, Donella H., Dennis L. Meadows, Jørgen Randers, and William W. Behrens III (Club of Rome). *The Limits to Growth.* New York: Universe Books, 1972; Daly, Herman. *Steady-State Economics.* San Francisco: Freeman, 1977; Schumacher, E. F. *Small Is Beautiful: Economics as if People Mattered.* New York: HarperCollins Publishers, 1973.

29. My own university's definition of sustainability incorporates the concept of limits: "Sustainability is the process of living within the limits of available physical, natural and social resources in ways that allow the living systems in which humans are embedded to thrive in perpetuity." See University of Alberta, Office of Sustainability. sustainability.ualberta.ca.

30. Daly, Herman E. "Toward Some Operational Principles of Sustainable Development." *Ecological Economics* 2(1990): 1–6.

31. *Our Common Future,* paragraph 27. Note, too, that David Griggs has reformulated this definition as follows: Sustainable development is "development that meets the needs of the present while safeguarding Earth's

life-support system, on which the welfare of current and future generations depends." Griggs, David, et al. "Sustainable Development Goals for People and Planet." *Nature* 495 (21 March 2013): 305–307, cited on page 306.

32. Ehrenfeld, John R. *Sustainability by Design: A Subversive Strategy for Transforming Our Culture.* New Haven, CT: Yale University Press, 2008, 6.

33. Weisman, Alan. *The World Without Us.* New York: Thomas Dunne Books, 2007. Weisman might be thought of as promoting an "ecocentric" viewpoint, which was expressed most powerfully in the 1960s and after by Arne Naess, whose concept of "deep ecology" holds that all living organisms have inherent value beyond their instrumental utility for humans. More prosaically, it's the idea that humans should be less self-centered.

34. Diamond, Jared. *Guns, Germs, and Steel: The Fates of Human Societies.* New York: Norton, 1999, 147.

35. Pre-industrial indigenous societies remain an inspiration for many sustainists, even if few people think that industrial society could actually "return" to these less impactful modes of existence in a literal way.

36. Heinberg, Richard. *Powerdown: Options and Actions for a Post-Carbon World.* Gabriola, BC: New Society Publishers, 2004.

37. Rifkin, *The Third Industrial Revolution,* 4–5.

38. Brown, Lester R. *Building a Sustainable Society.* New York: Norton, 1981, 278.

39. The distinction between sustainability and environmentalism becomes obvious in Shellenberger, Michael and Ted Nordhaus. "The Death of Environmentalism: Global Warming Politics in a Post-Environmental World." *Special Issue: Don't Fear the Reapers: On the Alleged Death of Environmentalism. Grist,* 13 January 2005. Online: http://grist.org/article/doe-reprint/. More recently, they have written a book on post-environmentalism: Shellenberger, Michael and Ted Nordhaus. *Break Through: From the Death of Environmentalism to the Politics of Possibility.* Boston: Houghton Mifflin, 2007. They offer a critique of the narrow parameters of environmentalism.

40. I have always felt uncomfortable with the term "Western," because of the inaccurate and unhelpful binaries that it implies. It will be used in this book simply as a shorthand designation for the industrial societies in Europe and North America.

Chapter 1

1. This is not to imply that there wasn't pollution before the nineteenth century. Numerous artisanal crafts contaminated waterways, and the burning of wood (and later coal) polluted urban centers in many parts of the world. See Thomas, Keith. *Man and the Natural World: Changing Attitudes in England, 1500–1800.* London: Allen Lane, 1983, 244–245.

2. Cronon, William. *Changes in the Land: Indians, Colonists, and the Ecology of New England.* New York: Hill & Wang, 1983, 11.
3. Mann, Charles C. *1493: Uncovering the New World Columbus Created.* New York: Knopf, 2011.
4. Thomas, *Man and the Natural World,* 193.
5. Williams, Michael. *Deforesting the Earth: From Prehistory to Global Crisis.* Chicago: University of Chicago Press, 2002, 269.
6. The Little Ice Age was most likely caused by a temporary decline in sunspot activity that produced diminished solar-radiation intensity, although some scholars believe that rapid ecological changes in the New World, during the early colonial period, caused or exacerbated the cooling process in the Atlantic world. The main point is that things got cold for a while, especially in Europe.
7. Diamond, *Collapse,* 11. See also, Tainter, Joseph. *The Collapse of Complex Societies.* Cambridge, UK: Cambridge University Press, 1988; reprint, 2003, off of whom Diamond builds much of his argument.
8. Grove, Richard H. *Green Imperialism: Colonial Expansion, Tropical Island Edens and the Origins of Environmentalism, 1600–1860.* Cambridge, UK: Cambridge University Press, 1995, 56: "Dr. Thomas Preston even told a Commons committee as late as 1791 that the decline of oak trees in England was 'not to be regretted for it is certain proof of national improvement and for Royal Navies countries yet barbarous are the right and proper nurseries.'" Grove notes that England turned to its American colonies for wood between 1690 and 1776.
9. See, for instance, Mckendrick, Neil, John Brewer, and J. H. Plumb. *The Birth of a Consumer Society: The Commercialization of Eighteenth-Century England.* Bloomington: Indiana University Press, 1982; reprint 1985.
10. Richardson, David. "Involuntary Migration in the Early Modern World." In *The Cambridge World History of Slavery.* Vol. 3. Eds. Bradley, Keith and Paul Cartledge. Cambridge, UK: Cambridge University Press, 2011, 583.
11. Crosby, Alfred W. *Ecological Imperialism: The Biological Expansion of Europe, 900–1900.* Cambridge, UK: Cambridge University Press, 1986.
12. For instance, Klooster, Wim. *Revolutions in the Atlantic World: A Comparative History.* New York: New York University Press, 2009; also Hunt, Lynn. *Inventing Human Rights: A History.* New York: Norton, 2007.
13. Lang, Helen S. *Aristotle's Physics and Its Medieval Varieties.* Albany: SUNY Press, 1992.
14. Obviously, I'm simplifying a complex subject here, since there were competing schools of thought and cosmologies at this time. But scholars began to agree on some fairly basic paradigmatic assumptions. See Shapin, Steven. *The Scientific Revolution.* Chicago: University of Chicago Press, 1996.

15. See Caradonna, Jeremy L. *The Enlightenment in Practice: Academic Prize Contests and Intellectual Culture in France, 1670–1794.* Ithaca, NY: Cornell University Press, 2012.

16. Thomas, *Man and the Natural World*, 15.

17. Worster, *Nature's Economy*, 36. Note that scholars disagree on the relationship between religion and environmental destruction. See my discussion of Lynn White Jr. in chapter 3.

18. Descartes, René. *Discourse on Method and Meditations on First Philosophy.* 4th ed. Trans. Donald. A. Cress. Indianapolis: Hackett, 1998, 35. Grober argues that Descartes's reader, Baruch Spinoza, departed from Descartes's anthropocentrism with his pantheism, which implied that humans were part of nature (or made from the same substance). Perhaps Spinoza deserves a greater place in the history of ecological thought. See Grober, *Sustainability*, 55.

19. Bacon discusses this theme in many of his works, but the line comes from "The Masculine Birth of Time," 1605.

20. Worster, *Nature's Economy*, 53. These are Worster's words to describe Smith.

21. Baron d'Holbach, Paul Henri Thiry. *Système de la nature, ou Des loix du monde physique et du monde moral.* Londres: M.-M. Rey, n.p., 1770.

22. Comte de Buffon, Georges-Louis Leclerc. *Histoire naturelle, générale et particulière, avec la description du Cabinet du Roi.* 36 vols. Paris: Impr. Royale, 1749–1788.

23. Linnaeus, Carl. *Systema Naturae.* London: Natural History Museum, 1991. Originally published in numerous multi-volume sets between 1735 and the 1790s.

24. Diderot, Denis and Jean le Rond d'Alembert, eds. *Encyclopédie, ou Dictionnaire raisonné des sciences, des arts et des métiers.* 28 vols. Paris, 1751–1772.

25. Grove, *Green Imperialism*, 50–51.

26. Worster, *Nature's Economy*, 2.

27. Worster, *Nature's Economy*, 7. Also, see White, Gilbert. *The Natural History of Selbourne.* London: White, 1789.

28. Much of the material for this and the following paragraph come from Caradonna, Jeremy L. "Conservationism *avant la lettre*? Public Essay Competitions on Forestry and Deforestation in Eighteenth-Century France." In *Invaluable Trees: Cultures of Nature, 1660–1830.* Eds. Laura Auricchio, Elizabeth Heckendorn, Cook, and Giulia Pacini. Oxford: Voltaire Foundation, 2012, 39–54.

29. Williams, *Deforesting the Earth*, 264.

30. Evelyn, John. *Sylva, or a Discourse of Forest-Trees and the Propagation of Timber in His Majesty's Dominions.* 1st ed. London: Allestry and Martyn, 1664. Note, too, that Evelyn was one of the first early moderns to write on urban air pollution. See his *Fumifugium, or, The Inconvenience of the Aer and*

Smoak of London Dissipated Together with some Remedies Humbly Proposed by J. E. esq. to His Sacred Majesty, and to the Parliament Now Assembled. London: W. Godbid, 1661.

31. Thomas, *Man and the Natural World*, 198. Also, Grove notes that both peasants and defenders of private property often opposed these conservation efforts. There were even riots in some places. Grove, *Green Imperialism*, 56-57.

32. See Grober, *Sustainability*, 64-70, cited on 70.

33. *Ordonnance sur le fait des Eaux et Forêts.* Paris: Chez P. Le Petit, 1669.

34. Whited, Tamara L. *Forests and Peasant Politics in Modern France.* New Haven, CT: Yale University Press, 2000, 22.

35. Whited, *Forests and Peasant Politics*, 23, 26; see the many sources on this subject cited in the notes of Caradonna, "Conservationism."

36. Grober, *Sustainability*, 75.

37. Carlowitz, Hans Carl von. *Sylvicultura oeconomica, oder haußwirthliche Nachricht und Naturmäßige Anweisung zur wilden Baum-Zucht.* Leipzig: Braun, 1713. As far as I know, there are only partial English translations of this work.

38. Grober, Ulrich. "Der Erfinder der Nachhaltigkeit." *Die Zeit* 48 (25 November 1999): 98. Online: http://www.zeit.de/1999/48/Der_Erfinder_der_Nachhaltigkeit. See also Grober. *Deep Roots: A Conceptual History of "Sustainability."* Berlin: Wissenschaftszentrum Berlin für Sozialforschung, 2007. Grober is the undisputed expert on Carlowitz.

39. Grober, *Deep Roots*, 17-18.

40. Grober, *Sustainability*, 83.

41. Carlowitz, *Sylvicultura oeconomica*, 105-106.

42. Grober, *Deep Roots*, 21-22; Grober, *Sustainability*, 116-118.

43. Grober, "Der Erfinder," n.p.

44. Grober, *Deep Roots*, 19.

45. Grober, *Deep Roots*, 19-20.

46. Labrousse, Ernst. *Esquisse du mouvement des prix et des revenus en France au XVIIIe siècle.* Vol. 2. Paris: Librairie Dalloz, 1933, 343-348.

47. This paragraph is a summary of Caradonna, "Conservationism." The essayists were participants in public writing competitions sponsored and judged by academies in Paris and the provinces.

48. See Diamond, *Collapse*, 277-306.

49. See Mann, Charles C. *1493: Uncovering the New World Columbus Created.* New York: Vintage, 2011.

50. Diamond, *Collapse*, 299-304.

51. Diamond, Collapse, 300-304; see also Richards, John F. *The Unending Frontier: An Environmental History of the Early Modern World.* Berkeley: University of California Press, 2006, 184-185.

52. Diamond, *Collapse*, 304.

53. Diamond, *Collapse*, 304; Iwamoto, Junichi. "The Development of Japanese Forestry." In *Forestry and the Forest Industry in Japan*. Ed. Yoshiya Iwai. Vancouver: UBC Press, 2002, 9.
54. Iwamoto, "The Development," 8.
55. Grove, *Green Imperialism*, 61.
56. Grove, *Green Imperialism*, 62, 387.
57. Grove, *Green Imperialism*, 6.
58. Grove, *Green Imperialism*, 15. See also 23.
59. Grove, *Green Imperialism*, 67, 155, 166.
60. Grove, *Green Imperialism*, 168.
61. Rifkin, *The Third Industrial Revolution*, 193.
62. Smith, Adam. *An Inquiry into the Nature and Causes of the Wealth of Nations*. London: Printed for W. Strahan and T. Cadel, 1776. Note that Bernard Mandeville's 1705 *The Fable of the Bees: Or, Private Vices, Public Benefits* (London: Printed for Sam Ballard, at the Blue-Ball, in Little-Britain) anticipated many of Smith's economic ideas, including the idea of a division of labor and the invisible hand. Both thinkers were interested in harnessing greed and self-interest for the good of the community.
63. Cited in Rifkin, *The Third Industrial Revolution*, 194.
64. See Caradonna, *The Enlightenment in Practice*, 170.
65. Brewer, Anthony. *The Making of the Classical Theory of Economic Growth*. London: Routledge, 2010, preface (n.p.).
66. Mokyr, Joel. *The Enlightened Economy: An Economic History of Britain, 1700–1850*. New Haven, CT: Yale Univ. Press, 2009, 13. On p. 487 he concludes that "the Enlightenment is the 600-pound gorilla in the room of modern economic growth."
67. Brewer, *The Making of the Classical Theory*, 3–4, 12–13.
68. See Caradonna, "Conservationism."
69. The Dutch, however, had already moved away from mercantilism in the seventeenth century. The British were headed in the same direction in the eighteenth century, although it wasn't until the following century that Britain created a truly deregulated economy.
70. Shovlin, John. *The Political Economy of Virtue: Luxury, Patriotism, and the Origins of the French Revolution*. Ithaca, NY: Cornell University Press, 2006.
71. Bibliothèque Municipale de Bordeaux, 828 (C) (essay #3, 1787).
72. The most recent work on the physiocrats shows how much resistance there was to their economic, legal, and political views. See Vardi, Liana. *The Physiocrats and the World of the Enlightenment*. New York: Cambridge University Press, 2012.
73. Rousseau, Jean-Jacques. *Discours sur les sciences et les arts*. Geneva: Barillot, 1750. Some of the same themes appear in *Émile, ou de l'éducation*, his famous treatise of 1762 on education.

74. Writers before Rousseau had suggested that humans were naturally equal, most notably Baruch Spinoza in his *Ethics*.
75. Rousseau, Jean-Jacques. *Discours sur l'origine et les fondements de l'inégalité parmi les hommes*, 1754.

Chapter 2

1. Mokyr, *The Enlightened Economy*, 489.
2. On narratives and historical consciousness, see White, Hayden. *The Content of the Form: Narrative Discourse and Historical Representation*. Baltimore: Johns Hopkins University Press, 1987.
3. On Enlightenment narratives, see Edelstein, Dan. *The Enlightenment: A Genealogy*. Chicago: University of Chicago Press, 2010.
4. Cited on the EPA's website, "The Causes of Climate Change," April 2013: http://www.epa.gov/climatechange/science/causes.html.
5. Sale, Kirkpatrick. *Rebels Against the Future: The Luddites and Their War on the Industrial Revolution: Lessons for the Computer Age*. Reading, MA: Addison-Wesley, 1995, 23.
6. Sale, *Rebels*, 199.
7. Ortiz, Isabel and Matthew Cummins. *Global Inequality: Beyond the Bottom Billion: A Rapid Review of Income Distribution in 141 Countries*. New York: UNICEF, 2011, 19.
8. Sachs, *Common Wealth*, 6, 247.
9. Sale, *Rebels*, 26–27.
10. Mokyr, *Enlightened Economy*, 100.
11. According to Donald Worster, England witnessed the enclosure (privatization) of 6.5 million acres of farmland in the late eighteenth and early nineteenth centuries. Worster, *Nature's Economy*, 13.
12. Sale, *Rebels*, 53.
13. Most of the material from this paragraph comes from Sale, *Rebels*, 27–55.
14. Sale, *Rebels*, 1–13. The quote comes from the "Author's Note."
15. Sale, *Rebels*, 16.
16. Munby, Lionel M., ed. *The Luddites and other Essays*. London: Micheal Katanka, 1971, 34; see also, in the same volume, Jenkinn, Aflred J. "Chartism and the Trade Unions," 76–77.
17. Munby, *The Luddites and other Essays*, 50.
18. Wordsworth, *Excursion*, Book 8, lines 105–109, 128–130, cited in Sale, *Rebels*, 54.
19. Worster, *Nature's Economy*, 82.
20. Transcendentalism refers to a philosophical movement in New England in the 1820s and 1830s associated with Bronson Alcott, Margaret Fuller, George Ripley, Henry David Thoreau, Ralph Waldo Emerson, and others,

which stressed the innate benevolence of humankind, and valued independence and self-suffuciency.

21. Thoreau, Henry David. *Walden; Or, Life in the Woods*. Boston: Ticknor and Fields, 1854.

22. Emerson, Ralph Waldo. "Self-Reliance." *Essays: First Series*. Boston, MA: [n.p.], 1841.

23. Thoreau, Henry David. "A Winter Walk." *The Dial* 4 (October 1843): 221–226; Thoreau, *The Maine Woods*. Boston: Ticknor and Fields, 1864. See also Worster, *Nature's Economy*, 58–63.

24. Steiguer, J.E. de. *The Origins of Modern Environmental Thought*. Tucson: University of Arizona Press, 2006, 8.

25. Nash, Roderick. *The Rights of Nature: A History of Environmental Ethics*. Madison: University of Wisconsin Press, 1989, 36.

26. Edwards, *Sustainability Revolution*, 12.

27. Engels, Friedrich. *The Condition of the Working-Class in England 1844*. Trans. Florence Kelley Wischnewetzky. London: Allen and Unwin, 1968, 20. This is a reprint of the 1845 edition with Engels's *English Dedication*.

28. Engels, *Condition*, 43.

29. Engels, *Condition*, 49–50.

30. Gottlieb, Robert. *Forcing the Spring: The Transformation of the American Environmental Movement*. Washington, DC: Island Press, 1993, 278.

31. Examples come later in this chapter and in chapter 4. Note that Mill was a political and philosophical liberal but not a socialist or backer of the far left, although he had respect for Charles Fourier and his ideas about utopian worker cooperatives.

32. Malthus, Thomas Robert. *An Essay on the Principle of Population: Or a View of Its Past and Present Effects on Human Happiness; with an Inquiry into Our Propsects Respecting the Future Removal or Mitigation of the Evils Which It Occasions*. 6th ed. London: John Murray, 1798; reprint 1826, I.I.16. Online: http://www.econlib.org/library/Malthus/malPlong1.html.

33. Malthus, *Essay*, I.II.1–I.II.9.

34. Malthus, *Essay*, I.II.8–I.II.11.

35. Wrigley, E. A. "The Limits to Growth: Malthus and the Classical Economists." *Population and Development Review*, 14 (1988): 30–48. See pages 34–35.

36. Brewer, *The Making of the Classical Theory of Economic Growth*, 4.

37. See chapter 4.

38. Mishan, *Economic Growth Debate*, 81.

39. Skidelsky, Robert and Edward Skidelsky. *How Much Is Enough? Money and the Good Life*. New York: Other Press, 2012, 53. Steiguer, *Origins*, 6.

40. Mokyr, *Enlightened Economy*, 5–6.

41. For discussions of ongoing debates about Malthus, his theories, and his relevance, see Sachs, *Common Wealth*, 73; Dryzek, *Politics of the Earth*, 28; and Bartlett, "Reflections on Sustainability," 15.
42. Diamond, *Collapse*, 311.
43. Diamond argues that, although ethnic tensions between Hutus and Tutsis certainly played a major role in the genocide, "it sill takes some explaining why a Hutu community would kill at least 5% of its [own] members in the absence of ethnic motives." Diamond argues that ethnic tensions were but one aspect of the genocidal violence. He contends that demographic pressures from scarce land, low food yields, and overpopulation created a civil conflict in which ethnic tensions became exacerbated. But the violence was ultimately sparked by a Malthusian demographic crisis. Diamond, *Collapse*, 319.
44. Meadows et al., *Limits to Growth: The 30-Year Update*, x–xi.
45. Ricardo, David. *On the Principles of Political Economy and Taxation*, in *The Works and Correspondence of David Ricardo*. Vol. 1. 3d ed. Ed. P. Sraffa. Cambridge, UK: Cambridge University Press, 1951, 120.
46. Wrigley, "The Limits to Growth," 40.
47. Brown, *Building a Sustainable Society*, 117–118.
48. Jevons, William Stanley. *The Coal Question; An Inquiry Concerning the Progress of the Nation, and the Probable Exhaustion of Our Coal-Mines*. 2d ed. London: Macmillan, 1866. Originally published in 1865. See chapter 7. This edition is unpaginated. Online: http://oll.libertyfund.org/?option=com_staticxt&staticfile=show.php%3Ftitle=317&Itemid=27.
49. Daly, *Steady-State Economics*, 121.
50. Jackson, Tim. *Prosperity Without Growth: Economics for a Finite Planet*. London: Earthscan, 2009, 54.
51. Wackernagel, Mathis and William E. Rees. "Perceptual and Structural Barriers to Investing in Natural Capital: Economics from an Ecological Footprint Perspective." *Ecological Economics*. 20.1 (January 1997): 20.
52. It is worth pointing out that Jevons speculated on the potential to harness solar power, which he considered a futuristic marvel and a threat to Britain's economic might: "Among the residual possibilities of unforeseen events, it is just possible that some day the sunbeams may be collected, or that some source of force now unknown may be detected. But such a discovery would simply destroy our peculiar industrial supremacy. The study of electricity has already been zealously cultivated on the Continent with this view." Jevons, *The Coal Question*, chapter 8.
53. See Steiguer, *Origins*, 7.
54. Mill, John Stuart. *Principles of Political Economy with Some of Their Applications to Social Philosophy*. 7th ed. London: Longmans, Green, 1909,

IV.6.1—IV.6.2; Originally published in 1848. Online: http://www.econ-lib.org/library/Mill/mlP61.html.

55. Mill, *Principles,* IV.6.4.
56. Mill, *Principles,* IV.6.5.
57. Mill, *Principles,* IV.6.9.
58. Mill, *Principles,* IV.6.6.
59. Skidelsky and Skidelsky, *How Much Is Enough,* 53–54.
60. Jackson, *Prosperity Without Growth,* 124.
61. Daly, *Steady-State Economics,* 18.
62. There are a few possible exceptions, however, which would include the fol-lowing: Thorstein Veblen's *The Theory of the Leisure Class* (1899), which deals with the economics of inequality and "conspicuous consumption," and Nicolas Pigou's *The Economics of Welfare* (1920), which argues that producers need to bear the full "cost" of production, including pollution, illness, and environmental degradation that stems from that production. Pigou suggested that taxes be levied on pollutive producers as a kind of corrective against unwanted external costs. See Veblen, Thorstein. *The Theory of the Leisure Class.* New York: Dover Publications, 1994; Pigou, Nicolas. *The Economics of Welfare.* London: Macmillan, 1932. Neoclassi-cal economics pushed the stationary state (and any proto-ecological eco-nomics) to the sidelines in the late nineteenth century, and the idea only resurfaced in a serious way in the late 1960s and 1970s. It would be hard to argue, for instance, that Alfred Marshall's *Principles of Economics* (1890) or Friedrich Hayek's *The Road to Serfdom* (1944) did much for ecological economics. See chapter 4 for a deeper analysis of ecological economics and its rejection of many of the fundamental tenets of neoclassical economics.
63. Darwin, Charles. *On the Origins of Species by Means of Natural Selection, or the Preservation of Favoured Races in the Struggle for Life.* London: Juhn Murray, 1859.
64. Worster, *Nature's Economy,* 192. Haeckel's book was called *Generelle Mor-phologie der Organismen* (Berlin: Reimer, 1866).
65. Even Darwin's *Descent of Man, and Selection in Relation to Sex* (1871), which is about human evolution, is not particularly interested in indus-trialism, although it was read by Victorians as a defense of the progress of European civilization. It is important to note, though, that Darwin was not a "social Darwinian."
66. Darwin, *On the Origins of Species,* chapter 3.
67. Sachs, Aaron. *The Humboldt Current: Nineteenth-Century Exploration and the Roots of American Environmentalism.* New York: Viking, 2006, 2.
68. Worster, *Nature's Economy,* 140–143.
69. Worster, *Nature's Economy,* 150–154.
70. Worster, *Nature's Economy,* 164.

71. Nash, *The Rights of Nature*, 22. Of course, the debate between Darwinian scientists and religious creationists lingers on in the twenty-first century.
72. Steiguer, *Origins*, 9.
73. "Marsh, George Perkins." In *The Encyclopedia of the Earth*. Online: http://www.eoearth.org/view/article/154491/.
74. Marsh, George Perkins. *Man and Nature, or, Physical Geography as Modified by Human Action*. Ed. David Lowenthal. Seattle: University of Washington Press, 2003, 37. Originally published in 1864.
75. Marsh, *Man and Nature*, 29–35. Marsh's book is fairly anthropocentric. He accepted that humans were the dominant power on Earth, but he believed that they could use this power to safeguard the environment. Marsh is often cited in the literature on environmental "stewardship"—the anthropocentirc idea that humans have a duty (and the capacity) to manage the natural environment.
76. "Marsh, George Perkins." *The Encyclopedia of the Earth*, n.p.
77. Marsh, *Man and Nature*, 39.
78. Marsh, *Man and Nature*, 43.
79. Nash, *The Rights of Nature*, 38.
80. Marsh, *The Rights of Nature*. xvii (editor's introduction).
81. Roosevelt often toed the line between Muir's and Pinchot's visions of national parks. Still, he was a more robust defender of conservationism than most of his predecessors and many of his successors. Here is Roosevelt on 2 December 1901 in a speech to Congress: "The fundamental idea of forestry is the perpetuation of forests by use. Forest protection is not an end in itself; it is a means to increase and sustain the resources of our country and the industries which depend upon them. The preservation of our forests is an imperative business necessity." Quoted in Gottlieb, *Forcing the Spring*, 23.
82. Quoted in Nash, *The Rights of Nature*, 40. See also Worster, *Nature's Economy*, 185.
83. Muir, John. *Our National Parks*. New York: Houghton Mifflin, 1901.
84. Muir, *Our National Parks*, chapter 1, n.p. Online: http://www.yosemite.ca.us/john_muir_writings/our_national_parks/chapter_1.html.
85. There were other problems with Pinchot's conservation, too—for instance, the fact that land conservation often included dispossessing Native Americans of traditionally used lands. A cynical interpretation of early conservationism would see it as an effort to preserve stolen land for the benefit of white colonial powers.
86. Righter, Robert W. *The Battle over Hetch Hetchy: America's Most Controversial Dam and the Birth of Modern Environmentalism*. Oxford: Oxford University Press, 2005, 68.
87. See Robinson, "Squaring the Circle? Some Thoughts on the Idea of Sustainable Development." *Ecological Economics* 48.4 (2004): 369–384."

The preservationist/conservationist distinction dates back to the earlier historical scholarship of Roderick Nash and Stephen Fox and has come under attack by some scholars who see Pinchot and Muir as a blend of both schools of thought. I think this critique is valid, but I believe conservationism and preservationism are nonetheless different land use strategies or conceptual traditions that continue to impact environmental policy in the North American and elsewhere.

88. Muir, John. "The Hetch-Hetchy Valley." *The Yosemite*. Chapter 12. New York: Century Company, 1912. Originally published in 1908 in the *Sierra Club Bulletin*. Also see Righter, *Battle*, 4; Steiguer, *Origins*, 12.
89. Rigther, *Battle*, 64–66.
90. Gottlieb, *Forcing the Spring*, 26.

Chapter 3

1. Jim Motavalli, Doug Moss, Brian C. Howard, and Karen Soucy, eds. *Green Living: The E Magazine Handbook for Living Lightly on the Earth*. London: Plume Book, 2005.
2. Williams, Raymond. "Ideas of Nature." In *Problems in Materialism and Culture: Selected Essays*. London: Verso, 1980, 67–85. Paul R. Ehrlich and John P. Holdren, in their essay from 1971 cited below, make a great case for an expanded definition of "environment." Of course, it was presumed throughout much of European history that humans, too, had a nature, but this nature was usually pitted in opposition to the effects of the civilizing process: Nature versus "the City." See Mumford, Lewis. *The City in History: Its Origins, Its Transformations, and Its Prospects*. New York: Harcourt, Brace & World, 1961. This is also the period in which James Lovelock and Lynn Margulis put forth the influential theory called the "gaia hypothesis" that conceptualized the Earth as a single ecological unit. Lovelock, James E. and Margulis, Lynn. "Atmosphetic Homeostasis by and for the Biosphere: The Gaia Hypothesis." *Tellus* Series A, 26.1–2 (1 February 1974): 2–10.
3. Worster, Donald. *Nature's Economy*, xiii-xiv; Worster, *The Wealth of Nature: Environmental History and the Ecological Imagination*. New York: Oxford University Press, 1994, chapter 13. Worster argues that ecologists no longer see ecosystems as stable but as chaotic and in flux. This change apparently began in the 1970s. He wonders aloud what the ethical implications are of this fundamental shift in thinking.
4. See, for instance, Dryzek, *The Politics of the Earth*, 3.
5. See Righter, *Battle over Hetch Hetchy*; Fox, Stephen R. *John Muir and His Legacy: The American Conservation Movement*. Boston: Little, Brown, 1981.

Note that the term "preservationism" declined in usage after Muir's death. "Conservationism" became the main term associated with environmental protection and policy. In a sense, Muir has been appropriated by conservationists, even though he saw himself as opposed to Pinchot's conception of conservation.

6. Leopold, Aldo. *A Sand County Almanac: And Sketches Here and There.* Oxford: Oxford University Press, 1949, 207, 204.

7. See Dryzek, *The Politics of the Earth*, 51.

8. The literature on the environmental movement is voluminous. A good place to start is with de Steiguer, *Origins*, and Gottlieb, *Forcing the Spring*; see also, Kline, Benjamin. *First along the River: A Brief History of the U.S. Environmental Movement.* 4th ed. Lanham, MD: Rowman & Littlefield, 2011; and Stoll, Steven. *U.S. Environmentalism since 1945: A Brief History with Documents.* Boston: Bedford/St. Martin's, 2007. For a more populist interpretation that rejects the conventional narrative that environmentalism began with Rachel Carson and social elites, see Montrie, Chad. *A People's History of Environmentalism in the United States.* New York: Continuum, 2011. Also, Schawb, Jim. *Deeper Shades of Green: The Rise of Blue-Collar and Minority Environmentalism in America.* San Francisco: Sierra Club Books, 1994.

9. See Mark Stoll's excellent "Rachael Carson's *Silent Spring*, A Book That Changed the World" on the Environment and Society Portal, run by the Rachel Carson Center for Environment and Society at the University of Munich. Online: http://www.environmentandsociety.org/exhibitions/silent-spring/silent-spring-international-best-seller.

10. Carson, Rachel. *Silent Spring.* Boston: Houghton Mifflin, 1962, 8.

11. Lytle, Mark. *The Gentle Subversive: Rachel Carson, Silent Spring, and the Rise of the Environmental* Movement. Oxford: Oxford University Press, 2007. See also the National Resources Defense Council's helpful summation of Carson's tribulations. Online: http://www.nrdc.org/health/pesticides/hcarson.asp.

12. Udall, Stewart L. *The Quiet Crisis.* New York: Holt, Rinehart and Winston, 1963. See also the later, updated editions of the book.

13. Ehrlich, Paul R. *The Population Bomb.* New York: Buccaneer Book, 1968; reprint 1995, 3.

14. In fairness to Ehrlich, the situation did seem fairly dire in the mid-1960s. To provide a bit of context to his warnings, there were a series of famines in this period: the Great Chinese Famine, the Biafran famine, and the Sahel drought killed many millions in Asia and Africa. Also, the rate of population growth did not begin to taper off until the 1970s, so that the world still has not doubled from 4 billion people (1974) to 8 billion. Also, Ehrlich was writing before the large increase in global food production that occurred

in the latter decades of the twentieth century. Ehrlich's response to all of this has been that humankind has merely delayed the catastrophe by, for instance, undermining long-term soil fertility through the use of ecologically destructive agricultural practices and synthetic chemicals.

15. Ehrlich, Paul R. and Holdren, John P. "Impact of Population Growth." *Science* n.s., 171.3977 (26 March 1971). 1212–1217.

16. White, Lynn, Jr. "The Historical Roots of Our Ecological Crisis." *Science* n.s. 155.3767 (10 March 1967), 1206.

17. Hardin, Garrett. "The Tragedy of the Commons." *Science* n.s. 162.3859 (13 December 1968): 1244.

18. Commoner, Barry. *The Closing Circle: Nature, Man, and Technology.* New York: Knopf, 1971, 138. See also the excellent secondary source on Commoner: Egan, Michael. *Barry Commoner and the Science of Survival: The Remaking of American Environmentalism.* Cambridge, MA: MIT, 2007.

19. Commoner, *The Closing Circle*, 29–42.

20. Jacobs, Jane. *The Death and Life of Great American Cities.* New York: Random House, 1961, 5.

21. Lefebvre, Henri. *Writings on Cities.* Eds. and trans. Eleonore Kofman and Elizabeth Lebas. Malden, MA: Blackwell, 1996.

22. "Declaration of the United Nations Conference on the Human Environment," Stockholm, 16 June 1972. Also known as the "Stockholm Declaration."

23. See Earth Day Network's website: www.earthday.org/about.

24. Some of those findings are summarized in "The Truth about Recycling." *The Economist*, 7 June 2007. Online: http://www.economist.com/node/9249262.

25. Gottlieb, *Forcing the Spring*, 96.

Chapter 4

1. See, for instance, Club of Rome, *Limits to Growth*, 49: "It makes no sense, therefore, to talk about growth with either unquestioning approval or unquestioning disapproval. Instead it is necessary to ask: Growth of what? For whom? At what cost? Paid by whom? What is the real need here, and what is the most direct and efficient way for those who have that need to satisfy it? How much is enough? What are the obligations to share? The answers to those questions can point the way toward a sufficient and equitable society."

2. I employ the term "economist" somewhat loosely in this chapter. Most of the scholars discussed in this chapter were, in fact, trained economists, but some were engineers, social scientists, or scientists who delved into

economics in their respective writings. Also, the term "ecological economics" was coined shortly after the period discussed in this chapter, and thus I use it here somewhat anachronistically. However, it is generally agreed that Kenneth Boulding, Howard T. Odum, E. J. Mishan, and Herman Daly developed this interdisciplinary field.

3. See Palley, Thomas. "Milton Friedman: The Great Laissez-Faire Partisan." *Economic & Political Weekly* 41.49 (9 December 2006): 5041–5043; Friedman, Milton. *Capitalism and Freedom*. Chicago: University of Chicago Press, 1962.

4. See, for instance, Keynes, John Maynard. *General Theory of Employment, Interest and Money*. New York: Macmillan, 1936; Hayek, Friedrich. *The Road to Serfdom*. London: Routledge, 1944.

5. Daly, Herman E. Farewell Speech to World Bank. 14 January 1994. Online: http://www.whirledbank.org/ourwords/daly.html.

6. Lovins, Amory. *Soft Energy Paths: Toward a Durable Peace*. Harmondsworth, UK: Penguin, 1977.

7. Stivers, *The Sustainable Society*, 77, also 52.

8. Daly, *Steady-State Economics*, 99. Recall, also, from my Introduction that Bill McKibben preferred the term "maturity" to sustainability.

9. See Keynes, John Maynard. "Economic Possibilities for Our Grandchildren." In *John Maynard Keynes, Essays in Persuasion*. New York: Norton, 1963. 358–373. Originally published in 1930; On Adam Smith see Brewer, *The Making of the Classical Theory of Economic Growth*, 137.

10. The response to the Club of Rome's critique of growth was swift and defensive. See Renshaw, Edward F. "Sustainable Growth: Accident or Design." *The Sustainable Society: Implications for Limited Growth*. Ed. D. C. Pirages. New York: Praeger, 1977, 132: "The initial reaction of most economists to the report by the Club of Rome on the limits to growth was not only skeptical but highly defensive of economic growth as being both desirable and necessary to achieve such worthwhile goals as a reduction of poverty and a cleaner environment." Renshaw's own work focuses on "adjusting to a no-growth economy."

11. Schumacher, *Small Is Beautiful*, 46.

12. Mishan, *The Cost of Economic Growth*, 17, 10.

13. Mishan, *The Cost of Economic Growth*, 29–30.

14. Mishan, *The Economic Growth Debate*; see also, Arndt, H. W. *The Rise and Fall of Economic Growth*, 1978.

15. Meadows, Donella H., Jørgen Randers, and Dennis L Meadows. *Limits to Growth: The 30-Year Update*. White River Junction, VT: Chelsea Green, x.

16. Club of Rome, *Limits to Growth: A Report for the Club of Rome's Project on the Predicament of Mankind*, 2d ed. New York: Universe Books, 1974, 25, 35. The report also introduced the term "carrying capacity," which

referred to the maximum human population that the Earth could safely sustain. They borrowed the term from shipbuilding.

17. Daly, Herman E. "Economics in a Full World." *Scientific American* 293.3 (2005): 100–110.

18. Daly, *Steady-State Economics*, 3.

19. See Mishan, *The Cost of Economic Growth*, 27.

20. Mishan, *The Cost of Economic Growth*, 64.

21. Daly, *Toward a Steady-State Economy*, 25.

22. Lovins. *Soft Energy Paths*, 6.

23. Hubbert, M. King. "Nuclear Energy and the Fossil Fuels." Paper presented at the Spring Meeting of the Southern District, American Petroleum Institute, Plaza Hotel, San Antonio, Texas, 7–9 March 1956. On peak oil, see Hirsch, Robert L., Bezdek, Roger, and Wendling, Robert. "Peaking of World Oil Production: Impacts, Mitigation & Risk Management." Science Applications International Corporation/US Department of Energy, February 2005. Online: www.netl.doe.gov/publications/others/pdf/Oil_Peaking_NETL.pdf; also see Richard Heinberg's many writings on the subject, including Heinberg, Richard. *The Party's Over: Oil, War and the Fate of Industrial Societies.* Gabriola, BC: New Society Publishers, 2003.

24. The term comes from Galbraith, John Kenneth. *The Affluent Society.* Boston: Houghton Mifflin, 1958. This book was influential for the ecological economists. It argues that since World War II, the United States has become wealthy in the private sector and poor in the public sector, exacerbating social inequalities and leaving limited funds for social and infrastructural development.

25. Mishan, *The Economic Growth Debate*, 29–30.

26. Daly, *Steady-State Economics*, 99–101. In this passage, Daly discusses the concept of uneconomic growth without actually using that term. See also Schumacher, *Small Is Beautiful*, 41, who scrutinizes conventional economic values: "This means that an activity can be economic though it plays hell with the environment, and that a competing activity, if at some cost it protects and conserves the environment, will be uneconomic."

27. Mishan, *The Cost of Economic Growth*, 65.

28. Daly, *Steady-State Economics*, 3.

29. Hirsch, Fred. *Social Limits to Growth.* Cambridge, MA: Harvard University Press, 1976, 4.

30. Pirages, *The Sustainable Society*, 8–11. Pages 10–11 deal with the idea of slow growth: "Sustainable growth is a difficult concept with which to deal, but it seems to be the best guide to the future that we have at present. It means economic growth that can be supported by physical and social environments for the foreseeable future. An ideal sustainable society would be one in which all energy would be derived from current solar income and

all nonrenewable resources would be recycled." Recall from the Introduction that Albert Bartlett calls "sustainable growth" an oxymoron.

31. Mishan, *The Cost of Economic Growth*, 32–35.
32. Schumacher, *Small Is Beautiful*, chapter 4. He began developing this idea in the mid-1960s.
33. Stivers, *The Sustainable Society*, 10.
34. Daly, *Steady-State Economics*, 148.
35. Daly, *Steady-State Economics*, 14; See also Mill, *Principles of Political Economy*, 1848. Mill released numerous later editions. Daly cites from the edition of 1857. For a good summary of the history of steady-state economics, see Anderson, Mark W. "Economics, Steady State." *Berkshire Encyclopedia of Sustainability*. Vol. 10, *The Future of Sustainability*. Great Barrington, MA: Berkshire, 2012, 78–85.
36. Daly, *Steady-State Economics*, 17.
37. Daly, *Steady-State Economics*, 53.
38. Daly, *Steady-State Economics*, 126.
39. Odum is cited below. See Holling, C. S. "Resilience and Stability of Ecological Systems." *Annual Review of Ecology and Systematics* 4 (November 1973): 1–23.
40. Mishan, *The Economic Growth Debate*, 32.
41. Mishan, *The Cost of Economic Growth*, 136.
42. Mishan, *The Cost of Economic Growth*, 82–83.
43. Mishan, *The Cost of Economic Growth*, 84. Mishan dates himself by arguing that air and water pollution have only "local" effects and thus the damage is easy to calculate. Mishan was writing in the mid-1960s, of course, before there was a well-formed understanding about the far-reaching effects of dispersed pollutants. He also tends to focus more on the social rather than the ecological consequences of pollution. See pages 83–84.
44. Boulding, Kenneth. "The Economics of the Coming Spaceship Earth." In Daly, *Toward a Steady-State Economy*, 127. This article is taken from his *Environmental Quality in a Growing Economy*. Ed. Henry Jarrett. Baltimore: Johns Hopkins University Press, 1966.
45. Boulding, "The Economics of the Coming Spaceship Earth," 125–127.
46. Georgescu-Roegen, Nicholas. *The Entropy Law and the Economic Process*. Cambridge, MA: Harvard University Press, 1971.
47. Odum, Howard T. and Elisabeth C. Odum. *Energy Basis for Man and Nature*. New York: McGraw-Hill Book Company, 1976, 5.
48. Daly, *Steady-State Economics*, 43.
49. Jackson, *Prosperity Without Growth*, 179.
50. Stivers, *The Sustainable Society*, 28.
51. Cited in Gore, Al. *The Future: Six Drivers of Global Change*. New York: Random House, 2013, 143.

52. Mishan, *The Cost of Economic Growth*, 39–40.
53. Daly, *Steady-State Economics*, 30: "We have no national accounting measures of either the cost or the benefits of growth, although we often treat GNP as a measure of benefits." Also in *Toward a Steady-State Economy* (pp. 4–5), Daly argues that there's a circular logic between full employment and GNP growth.
54. Daly, *Steady-State Economics*, 94.
55. Turner, *The Geography of Hope*, 278–279.
56. Daly, *Steady-State Economics*, 99.
57. Daly, *Steady-State Economics*, 9.
58. Mishan, *The Economic Growth Debate*, 32. As Schumacher put it, less prosaically: "The substance of man cannot be measured by Gross National Product." Schumacher, *Small Is Beautiful*, 18.
59. Boulding, "The Economics of the Coming Spaceship Earth," 127.
60. Stivers, *The Sustainable Society*, 31.
61. Schumacher, *Small Is Beautiful*, 17.
62. Schumacher, *Small Is Beautiful*, 18.
63. Lovins, *Soft Energy Paths*, 12–14, 26.
64. Odum and Odum, *Energy Basis for Man and Nature*, 63.
65. Odum and Odum, *Energy Basis for Man and Nature*, 20–32; also, see opening chapters of Schumacher, *Small Is Beautiful*.
66. Lovins, *Soft Energy Paths*, 38–39.
67. Daly, *Steady-State Economics*, 12.

Chapter 5

1. Mitchell Energy & Development Corp. organized three Woodlands Conferences (1975, 1977, and 1979) near Houston, Texas. This question was the topic of the first conference. See the conference proceedings: Coomer, James, C., ed. *Quest for a Sustainable Society*. New York: Pergamon, 1979, ix; Cleveland, Harlan, ed. *The Management of Sustainable Growth*. New York: Pergamon, 1981.
2. See, for instance, Meadows, Dennis L., ed. *Alternatives to Growth—I. A Search for Sustainable Futures*. Cambridge, MA: Ballinger, 1977; Coomer, *Quest for a Sustainable Society*; Cleveland, *The Management of Sustainable Growth*; Brown, *Building a Sustainable Society*; and many of the works cited in chapter 4.
3. Hamrin, Robert. "The Road to Qualitative Growth." In Cleveland, *The Management of Sustainable Growth*, 115.
4. Grober, *Sustainability*, 155.
5. Cleveland, Harlan. "We Changed Our Minds in the 1970s." In Cleveland, *The Management of Sustainable Growth*, 3.

6. World Commission on Environment and Development. *Report of the World Commission on Environment and Development: Our Common Future.* New York: United Nations, 1987, n.p. "Chairman's Foreword" by Brundtland; see also her memoirs: Brundtland, Gro Harlem. *Madam Prime Minister: A Life in Power and Politics.* New York: Farrar, Straus and Giroux, 2005.

7. United Nations, "Declaration of the United Nations Conference on the Human Environment," 1972. Many of the UN documents discussed here can be found in Rauschning, Dietrich, Katja Wiesbrock, and Martin Lailach. *Key Resolutions of the United Nations General Assembly: 1946-1996.* Cambridge, UK: Cambridge University Press, 1997.

8. International Union for the Conservation of Nature. *World Conservation Strategy: Living Resource Conservation for Sustainable Development.* Gland, Switzerland: IUCN, 1980, iv.

9. International Union for the Conservation of Nature. *World Conservation Strategy*, n.p., introduction.

10. World Commission on Environment and Development, *Our Common Future*, n.p., "Chariman's Foreword."

11. World Commission on Environment and Development, *Our Common Future*, n.p., "Chariman's Foreword."

12. World Commission on Environment and Development, *Our Common Future*, n.p., "From One Earth to One World."

13. Gordon, Lincoln. "Changing Growth Patterns and World Order." In Cleveland, *The Management of Sustainable Growth*, 276; also, Goodland, Robert, Herman E. Daly, Salah El Sarafy, and Bernd von Droste, eds. *Environmentally Sustainable Economic Development: Building on Brundtland.* Paris: UNESCO, 1991. The latter book recognized, even in the early 1990s, the need to distinguish between the economic growth of the West and the sustainable development of poorer countries. That is, the contributors to this volume did not want industrialized countries to turn development into merely another money-making scheme rooted in social and environmental exploitation.

14. Robinson, "Squaring the Circle?," 370. Donald Worster's *The Wealth of Nature* and Peter Victor's *Managing Without Growth* also address the conflict in values between sustainability and sustainable development.

15. Robinson, "Squaring the Circle?," 369-371.

16. Bartlett, "Reflections on Sustainability," 4-7.

17. International Union for the Conservation of Nature, *World Conservation Strategy*, Introduction, 32-33, 62.

18. World Commission on Environment and Development, *Our Common Future*, parts 1 and 2. The WCED backs growth in both the developed and developing world, and in that regard, it was at odds with much of the work in ecological economics. However, Herman Daly agreed that the developing world needed at least some economic growth to get out of severe

poverty and reduce local ecological destruction. See chapter 4 of this book. Here's *Our Common Future* on the same subject: "Poverty is not only an evil in itself, but sustainable development requires meeting the basic needs of all and extending to all the opportunity to fulfil their aspirations for a better life. A world in which poverty is endemic will always be prone to ecological and other catastrophes." See part 1.

19. Dryzek, *Politics of the Earth*, 147.

20. See the National Oceanic and Atmospheric Administration. "International Treaty Designed to Restore/Protect Ozone Layer Working, Say Scientists," 4 May 2006. Online: http://www.noaanews.noaa.gov/stories2006/s2624.htm.

21. By the 1960s, scientists had begun to connect the dots between industrial pollution and the warming of the planet. But then, in the 1970s, the long-term trend of global warming tapered off temporarily, leading some but certainly not all scientists to argue that the Earth was in a cooling period. It was quickly realized, however, that this "cooling" phase was little more than a temporary deviation from the warming trend. See Peterson, Thomas C., William M. Connolley, and John Fleck. "The Myth of the 1970s Global Cooling Scientific Consensus." *Bulletin of the American Meterological Society* 89 (2008): 1325–1337.

22. United Nations. "Rio Declaration on Environment and Development." Principle 3. See also Foo, Kim Boon. "The Rio Declaration and Its Influence on International Environmental Law." *Singapore Journal of Legal Studies* 1992(1992): 347–364.

23. See Beatley, Timothy. *Green Urbanism: Learning From European Cities*. Washington, DC: Island Press, 2000, 345–347; Robèrt, Karl-Henrik. *The Natural Step Story: Seeding a Quiet Revolution*. Gabriola Island, BC: New Society, 2002, 184.

24. United Nations, "Kyoto Protocol." For most signatories, the reductions were based on the GHG levels that existed in 1990. See Article 3. Also, even though the protocol discusses several different GHGs in the annex to the document, it's clear from the body of the text that carbon dioxide is the main target of the treaty.

25. United Nations, "Kyoto Protocol," Article 17. Emissions trading, or "cap and trade," is a market-based approach to reducing pollution (mainly air pollution but also water). Ordinarily, a government body sets a limit or cap on the total amount of pollution that a particular industry or set of industries can legally emit. The limit is divided up into "permits" or "carbon credits" (in the case of carbon trading) and allocated to businesses involved in the system. The permits correspond to the expected (and legally permissible) emissions of the firm. The permits are then bought and sold on a market. Firms that need to pollute above their allocated amount must

purchase more permits and those that pollute less can sell theirs. No new permits can enter the market until they have been spent at the end of a given period of time. Since the permits are purchased, there is an incentive for firms to pollute less and sell their excess permits at a profit. It is also meant to drive industry toward greener industrial practices. Firms that don't comply can be fined or penalized in various ways. The roots of cap and trade go back to the 1970s and 1980s, when the United States set up a trading scheme to reduce emissions of sulfur dioxide, which caused acid rain. The largest GHG trading system today is the European Union Emission Trading Scheme. In the Emission Trading Scheme, businesses that overpollute can either purchase permits from sellers or use the UN's Clean Development Mechanism to finance emissions reduction programs in the developing world and thus "offset" their emissions. This system creates a "hard cap," which stipulates that a firm or an industry must function beneath a defined emissions level. Elsewhere, there are "soft cap" emissions trading schemes that are "intensity based," which means that acceptable emissions levels are not absolute numbers but are rather tied to units of production. The soft cap scheme in Alberta allows industry to comply with emissions regulations in one of three ways: meeting targets, paying into a climate fund, or offsetting carbon emissions. Whether cap-and-trade systems are effective at reducing pollution remains a polarizing subject. Firms tend to like the market-based approach to emissions reduction. Yet the practice is frequently criticized as ineffectual, superficial, and easily manipulated. See, for instance, Shapiro, Mark. "Conning the Climate: Inside the Carbon-Trading Shell Game." *Harper's Magazine*, February 2010: 31–39.

26. See the essays in Coomer, *The Quest for a Sustainable Society*, and Cleveland, *The Management of Sustainable Growth.*

27. Brown, *Building a Sustainable Society*, 139–281, quote is on 278.

28. Brown, *Building a Sustainable Society*, 319–345.

29. Pearce, David, Anil Markandya, and Edward B. Barbier. *Blueprint 1: For a Green Economy*. London: Earthscan, 1989.

30. Some early programs in sustainability include the ones at Lund University (Sweden), Slippery Rock University (United States), and Appalachian State University (United States), although most programs at the BA, MA, and PhD levels were created after 2000.

31. Pezzey, John. *Sustainable Development Concepts*. Washington, DC: World Bank, 1992. Cited in the "Foreword" by Mohamad T. El Ashry.

32. Pezzey, *Sustainable Development Concepts*, 5.

33. Pezzey, *Sustainable Development Concepts*, 20, 60.

34. Carley, Michael and Ian Christie. *Managing Sustainable Development.* 2d ed. London: Earthscan, 2000, 27–28. The definition of sustainable

development presented in this work (p. 27) veers away from economic growth and focuses instead on environmental stewardship and social well-being: "Development is a process by which the members of a society increase their personal and institutional capacities to mobilize and manage resources to produce sustainable and justly distributed improvements in quality of life consistent with their own aspirations."

35. Carley and Christie, *Managing Sustainable Development*, 90, 111.

36. Bald eagles, and many other bird species, were poisoned through dichlorodiphenyltrichloroethane (DDT) biomagnification before that pesticide was banned. In 2013, the pesticide Clothianidin was determined to be the primary cause of the collapse of bee colonies in North America and elsewhere. Carrington, Damian. "Insecticide 'Unacceptable' Danger to Bees, Report Finds." *The Guardian* 16 January 2013. Online: http://www.guardian.co.uk/environment/2013/jan/16/insecticide-unacceptable-danger-bees/print.

37. Winter, Carl K. and Sarah F. Davis. "Scientific Status Summary: Organic Foods." *Journal of Food Science* 71: 9 (2006): R117–R124.

38. Winter and Davis, "Scientific Status Summary, R117.

39. See, for instance, this report from Grist.org: Laskawy, Tom. "Miracle Grow: Indian Farmers Smash Crop Yield Records Without GMOs." *Grist* 22 February 2013. Online: http://grist.org/food/miracle-grow-indian-farmers-smash-crop-yield-records-without-gmos/.

40. Mollison, Bill and David Holmgren. *Permaculture One: A Perennial Agriculture for Human Settlements*. Tyalgum, Australia: Tagari, 1981; also, Holmgren, *Permaculture: Principles and Pathways Beyond Sustainability*. Hepburn, Australia: Holmgren Design Services, 2002; reprint 2011.

41. Heinberg, Richard. *Peak Everything: Waking Up to the Century of Declines*. Gabriola Island, BC: New Society, 2007; reprint 2010, 59.

42. See Kai Kockerts, "The Fair Trade Story." Online: http://www.fairtrade.at/fileadmin/user_upload/PDFs/Fuer_Studierende/oikos_winner2_2005.pdf.

43. See the website of the Fairtrade Foundation: "Fairtrade Bucks Economic Trend with 19% Sales Growth," 25 February 2013. Online: http://www.fairtrade.org.uk/press_office/press_releases_and_statements/february_2013/fairtrade_bucks_economic_trend.aspx.

44. See "The Truth about Recycling."

45. Cited in "The Truth about Recycling."

46. Lovins, *Soft Energy Paths*; Brown, *Building a Sustainable Society*.

47. The three wind farms are the Altamont Pass Wind Farm, the Tehachapi Pass Wind Farm, and the San Gorgonio Pass Wind Farm.

48. There is a large literature on algae-based biofuels. For one example, see Kovacevic, V. and J. Wessler. "Cost-Effectiveness Analysis of Algae Energy Production in the EU." *Energy Policy* 38.10 (October 2010): 5749–5757.

49. See Tinsley, Dillard B. "Business Organizations in the Sustainable Society" in Coomer, *Quest for a Sustainable Society*, 164–167.
50. Brown, *Building a Sustainable Society*, 322–324.
51. Robèrt, *The Natural Step Story*, 65.
52. Hawken, Paul. *The Ecology of Commerce: A Declaration of Sustainability*. New York: Harper Business, 1993, xii.
53. Hawken, *The Ecology of Commerce*, xiii.
54. Hawken, *The Ecology of Commerce*, 10.
55. Hawken, *The Ecology of Commerce*, 1.
56. Hawken, *The Ecology of Commerce*, 144.
57. Hawken, Paul, Amory Lovins, and L. Hunter Lovins. *Natural Capitalism: Creating the Next Industrial Revolution*. New York: Little, Brown, 1999, 151.
58. Hawken, Paul, Amory Lovins, and L. Hunter Lovins. *Natural Capitalism: Creating the Next Industrial Revolution*. Boston: Little, Brown, 1999, 10–11.
59. Elkington, John. *Cannibals with Forks: The Triple Bottom Line of 21st Century Business*. Gabriola Island, BC: New Society, 1998, 70.
60. Elkington, *Cannibals with Forks*, 74–85.
61. See this 2013 report from *Environmental Leader*: "Companies Increasingly 'Pursue Triple Bottom Line,'" 1 May 2013. Online: http://www.environmentalleader.com/2013/05/01/companies-increasingly-pursue-triple-bottom-line/. Accessed in May of 2013.
62. See, for instance, The National Textile Center. "Annual Report: Strategic Sustainability and the Triple Bottom Line," November 2008. Online: http://www.ntcresearch.org/pdf-rpts/AnRp08/S06-AC01-A8.pdf.
63. Daly also wrote a work inspired by his time at the World Bank: Daly, Herman E. *Beyond Growth: The Economics of Sustainable Development*. Boston: Beacon Press, 1996.
64. Wackernagel, Mathis and William E. Rees. *Our Ecological Footprint: Reducing Human Impact on the Earth*. Gabriola Island, BC: New Society, 1996, 51–52.
65. Rees, William E. "Ecological Footprints and Appropriated Carrying Capacity: What Urban Economics Leaves Out." *Environment and Urbanization* 4.2 (October 1992): 121–130. See page 121.
66. Rees, "Ecological Footprints," 125–126.
67. Rees, "Ecological Footprints," 129.
68. See the website of the Global Footprint Network: http://www.footprintnetwork.org/en/index.php/GFN/. For an important critique of the ecological footprint, see Fiala, Nathan. "Measuring Sustainability: Why the Ecological Footprint Is Bad Economics and Bad Environmental Science." *Ecological Economics*. 67.4 (2008): 519–525. Fiala argues that EFA suffers from a fatal error in that it ignores the qualitative use of land and

assumes that less land use is better. He gives examples of how a society might be more sustainable if it used *more* land—for instance, in rotating crops rather than loading smaller areas of farmland with fertilizers and pesticides.

69. See Grober, *Sustainability*, 143.

70. The Bhopal disaster was a fatal gas leak in India that killed nearly 4,000 people and affected hundreds of thousands more. The plant that leaked the gas was owned by Union Carbide India Limited, which was a subsidiary of the Union Carbide Corporation. The plant manufactured pesticides. The Chernobyl disaster was a stupendous nuclear accident at the Chernobyl Nuclear Power Plant in the Ukraine. It exploded on 26 April 1986 and released large quantities of radioactive material into the atmosphere.

Chapter 6

1. Wackernagel, Mathis, et al. "Tracking the Ecological Over-Shoot of the Human Economy." *Proceedings of the National Academy of Sciences* 99.14 (9 July 2002): 9269.

2. GMOs are organisms that have been engineered at the genetic level. Genetically modified foods (GM foods) are foodstuffs that have been modified to resist disease, drought, pesticides, or herbicides. They raise a number of ecosystem concerns, and in most European countries, the law requires that GM foods be clearly labelled. The United States has not adopted this policy, despite the fact that nearly 90% of Americans want GM labelling. Over 10% of the world's farmland is now planted with GM crops. GM foods have *not* increased food yields and have created new vulnerabilities in the food supply. See Gore, *The Future*, 261–267.

3. "Agenda 21," section 8.41, called "Establishing Systems for Integrated Environmental and Economic Accounting."

4. For a more comprehensive discussion of new sustainability measurement tools, see Bosselmann, Klaus, Daniel S. Fogel, and J.B. Ruhl, eds. *Berkshire Encyclopedia of Sustainability*. Vol. 6, *Measurements, Indicators, and Research Methods for Sustainability*. Great Barrington, MA: Berkshire Publishing Group, 2012.

5. See Global Footprint Network: http://www.footprintnetwork.org/en/index.php/GFN/. The website also includes ecological and carbon footprint calculators.

6. See Braungart, Michael and William McDonough. *Cradle to Cradle: Remaking the Way We Make Things*. North Point Press, 2002.

7. See Redefining Progress: http://rprogress.org/sustainability_indicators/genuine_progress_indicator.htm

8. Anielski, Mark. *The Economics of Happiness: Building Genuine Wealth*. Gabriola, BC: New Society Publishers, 2007, 6 and xviii.

9. See Elkington, *Cannibals with Forks*, 1997. TBL is only one form of corporate ethics and alternative accounting but undoubtedly the most well known.

10. See the US Green Building Council: http://www.usgbc.org/. There are, however, other tools for measuring and promoting green building. In Canada, for instance, BOMA BESt is used as an environmental certification program for exisiting buildings.

11. See AASHE's website: https://stars.aashe.org/.

12. One critique of maximum sustained yield comes from the perspective of resilience. The idea of "resilience analysis, adaptive resource management, and adaptive governance" is conceived as an alternative to the notion of maximizing harvests, which assumes, rather problematically, that ecosystems are static. Adaptive resource management is more in tune with the vicissitudes and adaptations of a renewable resource (such as a fishery). See Walker, Brian, C. S. Holling, Stephen R. Carpenter, and Ann Kinzig. "Resilience, Adaptability and Transformability in Social-Ecological Systems." *Ecology and Society* 9.2 (2004): article 5. For adiscussion of EBFM, see Zolli, Andrew and Ann Marie Healy. *Resilience: Why Things Bounce Back*. New York: Free Press, 2012, 36–37.

13. See EcoLogo: http://www.ecologo.org/en/.

14. See http://www.organic.org/articles/showarticle/article-201.

15. See Fairtrade International: http://www.fairtrade.net/aims-of-fairtrade-standards.html.

16. See Ocean Wise: http://www.oceanwise.ca/about.

17. See the Marine Steward Council: http://www.msc.org/.

18. See the *Organic Trade Association's 2011 Organic Industry Survey*. See the summary online: http://www.ota.com/pics/documents/2011OrganicIndustrySurvey.pdf. On balance, the organic industry and organic certifying systems have been objects of a lot of criticism, whether it's the costs associated with getting certified or the confusing differences between products that are "made with organic ingredients," "organic," and "100% organic."

19. "LEED Buildings Grow by 14% Despite Market Crash." GreenBiz.com, 17 November 2010. Online: http://www.greenbiz.com/news/2010/11/17/leed-buildings-grow-14-percent-despite-market-crash.

20. Heinberg, *End of Growth*, 3–5, 16.

21. There are many problems with nuclear power. First, it produces a tiny fraction of the world's energy whereas it was billed in the mid-twentieth century as the solution to the world's energy woes. The IPCC says nuclear makes up 2% of the world's energy and the International Energy Agency

says 5.7%. See Intergovernmental Panel on Climate Change. *Renewable Energy Sources and Climate Change Mitigation*, 2011. Online: http://srren.ipcc-wg3.de/report; also, see International Energy Agency. *Key World Energy Statitics*, 2012. Online: https://www.iea.org/publications/freepublications/publication/kwes.pdf. Second, it is a nonrenewable resource since uranium is a finite resource. Third, most countries do not possess large uranium deposits, and therefore nuclear power prevents many countries from establishing energy independence. Fourth, only 31 of the world's countries have nuclear power plants as of 2013. Most countries either don't want or cannot support a nuclear power plant, and therefore this energy source is not applicable for much of the world. Fifth, nuclear power and nuclear weapons are linked since the ability to produce nuclear power also establishes the material basis and technological expertise to create nuclear weapons. Thus nuclear power, unlike renewables, is inextricably linked to destructive war-making potential. Sixth, nuclear waste is dangerously radioactive, difficult (or impossible) to store in a safe manner, and takes hundreds of thousands of years to lose its radioactivity. What to do with this waste is a huge debate in nearly every country that possesses a nuclear power plant. There are about 440 nuclear power plants today, and the waste from these plants is often stored on site in riduculously insufficient and aging metal bins. Some radioactive waste has been buried in caves or dumped in the world's oceans. Seventh, nuclear power plants are subject to catastrophic disasters, such as the 1979 accident at Three Mile Island in Pennsylvania, the 1986 Chernobyl disaster at the Chernobyl Nuclear Power Plant in the Ukraine, and the 2011 Fukushima Daiichi nuclear disaster in Japan. As a result of recent disasters, growth of new nuclear power plants has slowed way down, and some countries, including Germany, have decided to phase out their remaining plants over the next few decades. However, organizations such as the International Atomic Energy Association, which both oversees the world's nuclear industry and promotes nuclear power, have tried to use the threat of climate change as a basis from which to revive interest in nuclear power, since nuclear fission does not directly exacerbate anthropogenic climate change. (Dealing with spent-fuel and radiation exposure, however, does often require fossil fuels.) Still, it's hard to see how anyone could characterize nuclear power as a sustainable, safe, and renewable energy source, given the issues listed above.

22. Intergovernmental Panel on Climate Change, *Renewable Energy Sources and Climate Change Mitigation*, 10.

23. Sachs, *Common Wealth*, 96.

24. See Intergovernmental Panel on Climate Change, *Climate Change 2007: Synthesis Report*, 2007. Online: http://www.ipcc.ch/pdf/assessment-report/ar4/syr/ar4_syr.pdf.

25. Ocean energy refers to the "kinetic, thermal, and chemical energy of seawater, which can be transformed to provide electricity, thermal energy, or potable water." Intergovernmental Panel on Climate Change, *Renewable Energy Sources and Climate Change Mitigation*, 9.

26. It is important to distinguish between three terms that are often (and incorrectly) used interchangeably: **renewable energy** (or renewables), **sustainable energy**, and **energy independence**. Renewable energy refers to energy resources that can be continually replenished. Examples include trees (burning biomass), harnessing the energy from the sun, and corn (which can be turned into ethanol fuel). However, not all renewables are used sustainably. For instance, it is often claimed that corn ethanol takes more energy to produce than it yields. That is, it has a low EROI. Other biofuels, such as algae-based biodiesel and Barzilian sugarcane-based ethanol, might be more sustainable. Also, if forests are not replanted and managed sustainably, then they aren't really being renewed. A renewable resource could be quite pollutive, too, and therefore is not appropriately deemed sustainable. Sustainable energy is more of a qualitative assessment about how a renewable energy source is being used. Does it have an acceptable EROI? Is it safe and environmentally friendly? Does it meet current needs without compromising future generations? Does it promote peace, equality, and justice? Is it worth the associated costs (broadly defined)? Some kind of rating system must be applied to an energy source to determine if it's "sustainable." Finally, energy independence refers to the ability of a country to produce all of its own energy (automobile fuels, electricity sources, etc.). Energy indpendence can be accomplished either through sustainable energy or unsustainable and nonrenewable energy (at least in the short term). For instance, a country with lots of uranium and crude oil could claim to have energy independence, but neither of these resources are sustainable. Thus, within the sustainability movement, the goal is often to have renewable *and* sustainable energy in the hopes of achieving energy independence or something close to it. The problem is that there isn't a single, universally agreed upon metric for determining whether an energy source (or an energy installation) is sustainable, which leads to confusion and an abuse of the term.

27. Intergovernmental Panel on Climate Change, *Renewable Energy Sources and Climate Change Mitigation*, 9. Another excellent source on renewables is MacKay, David JC. *Sustainable Energy—Without Hot Air*. Cambridge, UK: UIT, 2008. Online: http://www.withouthotair.com/.

28. Intergovernmental Panel on Climate Change, *Renewable Energy Sources and Climate Change Mitigation*, 9.

29. See Denmark's informative website, which claims that the country will be free of fossil fuels by 2050. Denmark's energy is managed by State of Green, a public–private partnership between the government and

energy producers. State of Green: http://www.stateofgreen.com/en/ Intelligent-Energy.

30. Turner, *Geography of Hope*, 35.

31. See Germany's own energy report for 2011, Federal Ministry for the Environment, Nature Conservation and Nuclear Safety. *Development of Renewable Energy Sources in Germany 2011*, 2011–2012. Online: http://www.erneuerbare-energien.de/fileadmin/Daten_EE/Bilder_ Startseite/Bilder_Datenservice/PDFs__XLS/20130110_EEiZIU_E_ PPT_2011_FIN.pdf.

32. Turner, *Geography of Hope*, 309–312.

33. See page 2 of the UN's "Factsheet: The Need for Mitigation" for statistics on global emissions by sector: http://unfccc.int/files/press/background-ers/application/pdf/press_factsh_mitigation.pdf.

34. US Department of Energy. *Building Technologies Program*. Washington, DC: US Department of Energy, Energy Efficiency and Renewable Energy, 2008.

35. See the US Green Building Council's LEED website: http://www.usgbc.org/leed.

36. See the US Green Building Council's LEED website: http://www.usgbc.org/leed.

37. This is Navigant Research's industry projection, as discussed by Business Wire: "Green Building Materials Will Reach $254 Billion in Annual Market Value by 2020, Forecasts Navigant Research," 2 May 2013. Online: http://www.marketwatch.com/story/green-building-materials-will-reach-254-billion-in-annual-market-value-by-2020-forecasts-navigant-research-2013-05-02.

38. "German Solar Power Installations at Record High in 2012." Reuters, 5 January 2013. Online: http://www.reuters.com/article/2013/01/05/us-germany-solar-idUSBRE90406C20130105.

39. Federal Ministry for the Environment, Nature Conservation and Nuclear Safety, *Development of Renewable Energy Sources in Germany 2012*, 8: http://www.erneuerbare-energien.de/fileadmin/Daten_EE/Dokumente__PDFs_/ee_in_zahlen_ppt_en_bf.pdf.

40. Cogeneration or combined heat and power ordinarily refers to the simultaneous generation of heat (for heating buildings) and electricity from power stations. Denmark, for instance, has innovative interconnected industrial districts in which electricity-generating biomass power plants channel residual heat to neighboring buildings. "District heating" essentially puts to use thermal heat that would otherwise be released into the air and wasted.

41. Senick, Jennifer. "Green Building Benefits: By the Numbers." Rutgers Center for Green Building, 2011. Online: http://rcgb.rutgers.edu/uploaded_documents/Green.pdf.

42. See, for instance, United Nations Human Settlements Programme. *State of the World's Cities Report 2008/2009: Harmonious Cities*. London: UN-HABITAT, 2008.

43. Hallsmith, Gwendolyn and Bernard Lietaer. *Creating Wealth: Growing Local Economies with Local Currencies*. Gabriola, BC: New Society Publishers, 2011.

44. The data from this paragraph comes from Zovanyi, Gabor. *The No-Growth Imperative: Creating Sustainable Communities under Ecological Limits to Growth*. New York: Routledge, 2013, 12.

45. Zovanyi, *The No-Growth Imperative*, 12.

46. Hawken et al., *Natural Capitalism*, 22.

47. Zovanyi, *No-Growth Imperative*, 11–12.

48. See Wackernagel and Rees, *Our Ecological Footprint*.

49. Turner, *Geography of Hope*, 247–251. Note that New Urbanism has been criticized in some places for creating expensive niche neighborhoods that aren't always well integrated into surrounding areas.

50. Beatley, *Green Urbanism*, 6–8.

51. Beatley, *Green Urbanism*, 16–22.

52. Gehl, Jan. *Cities for People*. Washington, DC: Island Press, 2010, 7.

53. Gehl, *Cities for People*, 7.

54. See, for instance, this report by the Univeristy of Wisconsin Madison Urban and Regional Planning Department on "Urban Agriculture": http://urpl.wisc.edu/ecoplan/content/lit_urbanag.pdf.

55. Zovanyi, *No-Growth Imperative*, 60.

56. See Fodor, Eben. *Better Not Bigger*. Stony Creek, CT: New Society, 1999.

57. Florida, Richard. "Cities and the Creative Class." *City & Community* 2.1 (March 2003): 8.

58. Kagan, Sacha and Julia Hahn. "Creative Cities and (Un)Sustainability: From Creative Class to Sustainable Creative Cities." *Culture and Local Governance* 3.1–2 (2011): 11–27; See also Jones, Alison. "Selbstbestimmtes Leben: Hamburg's Rote Flora and the Roots of Autonomie in Twentieth-Century Germany." MA Thesis. University of Alberta, 2013.

59. See page 2 of the UN's energy report, "Fact Sheet: The Need for Mitigation": http://unfccc.int/files/press/backgrounders/application/pdf/press_factsh_mitigation.pdf. Here's the complete end-use emissions statistics by sector: fossil fuel supply (5%), waste (3%), power supply (21%), industry (19%), forestry (17%), agriculture (14%), transport (13%), and buildings (8%).

60. These figures are taken from Hawken et al., *Natural Capitalism*, 22–23.

61. See the Surface Transportation Policy Project. *Factsheet on Transportation and Climate Change*. Online: http://www.transact.org/library/factsheets/equity.asp.

62. Gehl, *Cities for People*, addresses this subject throughout.

63. See, for example, this report from Bloomberg.com: Downing, Louise. "Airlines Prepare to Take Off on Fuel Made from Algae, Wood Chips," 6 July 2011. Online: http://www.bloomberg.com/news/2011-07-07/airlines-prepare-to-take-off-on-fuel-made-from-algae-wood-chips.html. Also see from *Forbes*: Woody, Todd. "The U.S. Military's Great Green Gamble Spurs Biofuel Startups," 24 September 2012. Online: http://www.forbes.com/sites/toddwoody/2012/09/06/the-u-s-militarys-great-green-gamble-spurs-biofuel-startups/.

64. See AASHE at http://www.aashe.org/.

65. The Talloires Declaration: http://www.ulsf.org/programs_talloires.html.

66. See Columbia's program: http://gsas.columbia.edu/content/academic-programs/sustainable-development.

67. Rees, William E. "Impeding Sustainability? The Ecological Footprint of Higher Education?" *Planning for Higher Education* 31.3 (2003): 88–98.

68. Jackson, *Prosperity Without Growth*, 13.

69. Skidelsky and Skidelsky, *How Much Is Enough?*, 127.

70. Victor, *Managing Without Growth*, 22.

71. Heinberg, *The End of Growth*, 7, 189.

72. These are all themes in the work of Jackson, Victor, Heinberg, and other sustainability economists.

73. There is a growing literature of road maps, blue prints, and manuels for transitioning from industrial unsustainability to sustainable economies and communities. See, for instance, Lewis, Michael and Pat Conaty. *The Resilience Imperative: Cooperative Transitions to a Steady State Economy*. Gabriola, BC: New Society Publishers, 2012; Ehrenfeld, *Sustainability by Design*; Johnson, Huey D. *Green Plans: Greenprint for Sustainability*. Lincoln: University of Nebraska Press, 2008; Schendler, Auden. *Getting Green Done: Hard Truths from the Front Lines of the Sustainability Revolution*. New York: PublicAffairs, 2009; and Rob Hopkins' work on transition towns, cited below. Other such works are discussed in chapter 7.

74. Stern, Nicholas. "Stern Review: The Economics of Climate Change. Executive Summary." London: HM Treausry, 2006; see also, Jowit, Juliette and Patrick Wintour. "Cost of Tackling Climate Change has Doubled, Warns Stern." *The Guardian*, 26 June 2008. Online: http://www.guardian.co.uk/environment/2008/jun/26/climatechange.scienceofclimatechange. The idea, in the report, was to stabilize the parts per milion of CO_2 at 550. Stern, unlike many ecological economists, continues to pine for growth. But what was so interesting about his report is that it made clear that climate change will prevent future economic growth. Even for growthists,

the message was clear: there can be no prosperous long-term economy if climate change goes ignored. Inaction is costlier than the costs of preventative action.

75. See McKibben, Bill. *Deep Economy: The Wealth of Communities and the Durable Future.* New York: Henry Holt, 2007.

76. Jackson, *Prosperity Without Growth,* 200; Rifkin, *Third Industrial Revolution,* 107.

77. Easterlin, Richard. "Does Economic Growth Improve the Human Lot? Some Empirical Evidence." *Nations and Households in Economic Growth: Essays in Honor of Moses Abramowitz.* Eds. Paul A. David and Melvin W. Reder. New York: Academic Press, 1974, 89–125; Anielski, *The Economics of Happiness*; Layard, Richard. *Happiness: Lessons from a New Science.* New York: Penguin, 2006; Stiglitz, Joseph E., Amartya Sen, and Jean-Paul Fitoussi. *Report by the Commission on the Measurement of Economic Performance and Social Progress,* 2009. Online: http://www.neweconomics. org/. Note that Mishan, *Economic Growth Debate*; Victor, *Managing Without Growth,* 2008; and Jackson, *Prosperity Without Growth* also discuss the economics of happiness.

78. Anielski, *The Economics of Happiness,* 220; Happy Planet Index. *The (Un) Happy Planet Index.* Online: http://www.happyplanetindex.org/.

79. Skidelsky and Skidelsky, *How Much Is Enough?,* 105–113.

80. Rees, William E. "True Cost Economics." *Berkshire Encyclopedia of Sustainability.* Vol. 2, *The Business of Sustainability.* Eds. Chris Laszlo, Karen Christensen, and Daniel Fogel. Great Barrington, MA: Berkshire Publishing Group, 2010, 468; also, Victor states: "If ecosystem services were actually paid for, in terms of their value contribution to the global economy, the global price system would be very different from what it is today." Victor, *Managing Without Growth,* 42.

81. Rees, "True Cost Economics," 469.

82. This estimate comes from the National Resource Defense Council's 2012 report on fossil fuel subsidies. Online: http://endfossilfuelsubsidies.org/ files/2012/05/fossilfuelsubsidies_report-nrdc.pdf.

83. See the Rocky Mountain Institue: http://www.rmi.org/Walmartsfleet- operations.

84. See Green Analytics: http://greenanalytics.ca/index.php.

85. See The Green Life Online: http://site.thegreenlifeonline.org/green- wash101/.

86. See Southern Energy Management: http://www.southern-energy.com.

87. Jones, Van. *The Green-Collar Economy: How One Solution Can Fix Our Two Biggest Problems.* New York: HarperOne, 2008. n.p. (opening pages), 5.

88. Federal Ministry for the Environment, Nature Conservation and Nuclear Safety. *Renewable Energy Sources in Figures: National and*

International Development. 2011, 36. Online: http://www.erneuerbare-energien.de/fileadmin/ee-import/files/english/pdf/application/pdf/broschuere_ee_zahlen_en_bf.pdf.

89. Turner, *Geography of Hope*, 288.
90. Turner, *Geography of Hope*, 380–388; for Kiva, see http://www.kiva.org/.
91. See Hopkins, Rob. *The Transition Companion: Making Your Community More Resilient in Uncertain Times.* White River Junction, VT.: Chelsea Green Publishing, 2011, 288–289.
92. Even when investment structures remain the same, an ethical and ecological consciousness can change investment. For instance, the "divestment" movement has gained many adherents over the past couple of years. A growing number of entities (often universities and cities) are getting rid of their financial investments in fossil fuels and related industries in an effort to green their financial practices.
93. See the Capital Institute: http://www.capitalinstitute.org/about.
94. Cardwell, Diane. "For Solazyme, a Side Trip on the Way to Clean Fuel." *The New York Times,* 22 June 2013. Cardwell cites "the Cleantech Group's i3 Platform, a proprietary database" as her source.
95. Hawken, *Ecology of Commerce*, 1.
96. Stren, Richard E. and Mario Polèse. "Understanding the New Sociocultural Dynamics of Cities: Comparative Urban Policy in a Global Context." In *The Social Sustainability of Cities: Diversity and the Management of Change.* Eds. Mario Polèse and Richard E. Stren. Toronto: University of Toronto Press, 2000, 3.
97. Harris, Jonathan M. and Neva R. Goodwin. "Volume Introduction." In *A Survey of Sustainable Development: Social and Economic Dimensions.* Eds. Jonathan M. Harris, Timothy A. Wise, Kevin P. Gallagher, and Neva R. Goodwin. Washington, DC: Island Press, 2001, xxvii.
98. Harris and Goodwin, "Volume Introduction," xxix.
99. Gehl, *Cities for People*, 109.
100. See Cavanagh, J. and J. Mander, eds. *Alternatives to Economic Globalization: A Better World is Possible.* 2d ed. San Francisco: Barret-Koehler Publishers, 2004.
101. Owusu, K. and F. Ng'ambi. *Structural Damage: The Causes and Consequences of Malawi's Food Crisis.* Ed. Mark Ellis-Jones. London: World Development Movement, 2002, 20; Turner, *Geography of Hope*, 384.
102. See Daly, *Beyond Growth*, 5.
103. Harris and Goodwin, "Volume Introduction," xxxii.
104. Griggs et al., "Sustainable Development Goals for People and Planet," 306. The authors also propose updated MDGs. However, this proposed framework says little about creating economic equilibrium or transitioning away from a growth-centered economic system.

105. Sachs, *Common Wealth*, 131.
106. For instance, the UN believes that drinking water targets are on pace to be met by 2015, but scholars question the ability for sanitation goals to be met on time. Around 2.5 billion people lack access to sufficient sanitation. See Beavis, Sara G. "Water." In *Berkshire Encyclopedia of Sustainability*. Vol. 10, *The Future of Sustainability*. Ed. Daniel E. Vasey et al. Great Barrington, MA: Berkshire Publishing Group, 2012, 235.
107. Sachs, *Common Wealth*, 239–240.
108. UN MDGs: http://www.un.org/millenniumgoals/aids.shtml.
109. Diamond, *Collapse*, 497.
110. Diamond, *Collapse*, 329.
111. Daly, *Beyond Growth*, 31.
112. Jones, *Green-Collar Economy*, 56–57, quote on 58.
113. Hawken et al., *Natural Capitalism*, 23.
114. Robert, *Natural Step Story*, 26.
115. Smith, Alisa and J. B. Mackinnon. *The 100-Mile Diet: A Year of Local Eating*. Toronto: Random House Canada, 2007.
116. US Department of Agriculture. *Local Food Systems*, 2010, iii. Online: http://permanent.access.gpo.gov/lps125302/ERR97.pdf.
117. Hopkins, *The Transition Companion*, 44–60; see also, James, Sarah L., and Torbjorn Lahti. *The Natural Step for Communities: How Cities and Towns can Change to Sustainable Practices*. Gabriola, BC: New Society, 2009.
118. *Local Food Systems*, iii.
119. See, for instance, the data put forth by Baroni, L., et al. "Evaluating the Environmental Impact of Various Dietary Patterns Combined with Different Food Production Systems." *European Journal of Clinical Nutrition* 61(2006): 279–286. To sum up, organic veganism has the "smallest environmental impact." See page 5.
120. See the European Commission: http://ec.europa.eu/environment/eussd/escp_en.htm; and the City of Vancouver: http://vancouver.ca/files/cov/Greenest-city-action-plan.pdf.
121. Rifkin, Third Industrial Revolution, 42.
122. Nahi, Paul. "Government Subsidies: Silent Killer of Renewable Energy." *Forbes*, 14 February 2013.
123. See www.resilience.org and C.S. Holling's many writings on resilience.

Chapter 7

1. For a fuller exploration of these questions, see Goldblatt, David. *Sustainable Energy Consumption and Society: Personal, Technological, or Social Change?* Dordrecht: Springer, 2005.

2. Hallsmith, Gwendolyn and Bernard Lietaer. *Creating Wealth: Growing Local Economies with Local Currencies.* Gabriola Island, BC: New Society Publishers, 2011, 10.

3. Lowe, Ian. "Shaping a Sustainable Future—An Outline of the Transition." *Civil Engineering and Environmental Systems* 25.4 (December 2008): 247–254; Boyd, David R. *Sustainability Within a Generation: A New Vision for Canada.* Vancouver: David Suzuki Foundation, 2004.

4. There are different schools of thought on the green economy. Ecological modernization theory suggests that the market economy needs to "modernize" by adopting ecological principles and processes. Anti-growthists emphasize to need to reduce consumption, relocalize, and curb economic growth. Ecological economists want to reconceptualize capitalism altogether. But these various approaches share much in common.

5. See Schendler, Auden. *Getting Green Done: Hard Truths from the Front Lines of the Sustainability Revolution.* New York: PublicAffairs, 2010.

6. Ehrenfeld, *Sustainability by Design,* xix.

7. The uncritical love of growth is not only found in neoclassical economics and the business world but is central to governmental policymaking. In the words of Holmgren, "Governments do not generally support major social changes away from addictive consumption, even though the social and environmental benefits would be great, because the growth economy is inextricably tied (that is, addicted) to dysfunctional over-consumption." Holmgren, *Permaculture,* 113.

8. Reagan had already begun to chip away at financial regulations in the 1980s, but Clinton continued and even strengthened many of the neoliberal policies of the Republicans who preceded him. Likewise, in the United Kingdom, Tony Blair and Gordon Brown furthered some of the policies of deregulation and privatization begun by Thatcher.

9. Mason, Paul. *Meltdown: The End of the Age of Greed.* New York: Verso, 2010, 56–58, quote on 58.

10. Hawken, Paul. *Blessed Unrest: How the Largest Movement in the World Came into Being and Why No One Saw It Coming.* New York: Viking, 2007, 132.

11. Canada certainly felt the effects of the meltdown. Job losses occurred and some banks, such as the Canadian Imperial Bank of Commerce took heavy losses, but there weren't banking collapses as there were in the United States. Tighter banking regulations were generally cited as the reason that Canada was able to avoid the worst effects of the meltdown.

12. See chapters 4 and 6. On the issue of debt, see Mason, *Meltdown,* and Dienst, Richard. *The Bonds of Debt.* New York: Verso, 2011.

13. The quote is taken from an unpaginated blurb in Zovyani, *The No-Growth Imperative.*

14. See Anderson, "Economics, Steady State."

15. See Zovyani, *The No-Growth Imperative*, 182, and the ecological economists discussed in chapter 6.

16. The books by Heinberg, Lewis and Conaty, Fodor, and Hopkins are cited in previous chapters and can be referenced in the bibliography. Trainer's works are known primarily in Australia. See Trainer, Ted. *The Transition: To a Sustainable and Just World*. Canterbury, Australia: Envirobook, 2010.

17. Zehner, Ozzie. *Green Illusions: The Dirty Secrets of Clean Energy and the Future of Environmentalism*. Lincoln: University of Nebraska Press, 2012; Trainer, Ted. *Renewable Energy Cannot Sustain a Consumer Society*. Dordrecht: Springer, 2007.

18. A good primer on resilience is Zolli, Andrew and Ann Marie Healy. *Resilience: Why Things Bounce Back*. New York: Free Press, 2012. Although the authors (p. 21) see resilience and sustainability as contrasting approaches to ecological problems, I see both as part of the same broader discourse. Only a caricatured conception of sustainability-as-stasis would necessitate the distinction. Resilience should be considered a helpful addition to the sustainability movement.

19. Brown, Lester R. *Plan B 2.0: Rescuing a Planet under Stress and a Civilization in Trouble*. New York: Norton, 2006, 109. Brown cites US Geological Survey data to argue that the world could run out of lead reserves by 2024, tin by 2026, copper by 2031, iron ore by 2070, and bauxite by 2075.

20. Walker et al., "Resilience, Adaptability and Transformability in Social-Ecological Systems."

21. Lewis and Conaty, *The Resilience Imperative*, 19–20.

22. See resilience.org; http://www.postcarbon.org/; Intergovernmental Panel on Climate Change, *The Synthesis Report of the Fifth Assessment Report*; and Gore, *The Future*, 294, who cites a 2012 report from the World Bank.

23. Jones, *Green-Collar Economy*, 94.

24. Gore, *The Future*, 303.

25. This is a central theme in Sachs, *Common Wealth*.

26. Mason, *Meltdown*, chapters 1–6.

27. Gore, *The Future*, 303.

28. Rees, William E. "Thinking 'Resilience.'" In *The Post-Carbon Reader: Managing the 21st Century's Sustainability Crisis*. Eds. R. Heinberg and D. Lerch. Healdsburg, CA: Watershed Media, 2010, 29–30.

29. Walker et al., "Resilience, Adaptability and Transformability in Social-Ecological Systems."

30. Rees, "Thinking 'Resilience,'" 31.

31. See Griggs et al., "Sustainable Development Goals for People and Planet," 305–307; Postell, Sandra. "Water: Adapting to a New Normal." In *The Post-Carbon Reader: Managing the 21st Century's Sustainability Crisis*. Eds. R. Heinberg and D. Lerch. Watershed Media, 2010, 93.

32. See Hansen, James, Makiko Sato Reto Ruedy Ken Lo, David W. Lea and Martin Medina-Elizade. "Global Temperature Change." *PNAS* 103.39 (26 September 2006): 14288–14293; Flannery, Tim. *The Weather Makers: How We Are Changing the Climate and What It Means for Life on Earth*. New York: HarperCollins, 2006; *Climate Change and Biodiversity*. Eds. T. E. Lovejoy and L. Hannah. New Haven, CT: Yale University Press, 2006.

33. Crist, Eileen. "Beyond the Climate Crisis: A Critique of Climate Change Discourse." *Telos* 141 (Winter 2007), 33.

34. Crist, "Beyond the Climate Crisis," 40.

35. Crist, "Beyond the Climate Crisis," 36.

36. Trainer, Ted. "Can Renewable Energy Sustain Consumer Societies? A Negative Case." Simplicity Institute Report 12e, 2012, 15.

37. Holmgren, *Permaculture*.

38. TerraChoice. *The Sins of Greenwashing: Home and Family Edition*. 2010. Online: http://sinsofgreenwashing.org/index.html.

39. There are now many organizations and websites that track and fight green-washing. Many of those resources are discussed in an article in the *Globe and Mail* from 2012: Batista, Candice. "How Green Is It? How to Sort Out the Environmental Hype." *Globe and Mail*, 20 April 2012. Online: http://www.theglobeandmail.com/life/home-and-garden/spring-cleaning/how-green-is-it-how-to-sort-out-the-environmental-hype/article4101512/.

40. Monbiot, George. "The Denial Industry." *The Guardian*, 19 September 2006. See also Monbiot, George. *Heat: How to Stop the Planet from Burning*. New York: Random House, 2009.

41. Zehner, *Green Illusions*, 157.

42. Oreskes, Naomi and Erik M. Conway. *Merchants of Doubt: How a Handful of Scientists Obscured the Truth on Issues from Tobacco Smoke to Global Warming*. New York: Bloomsbury, 2010.

43. Zehner, *Green Illusions*, 157–158.

44. See the report *Who's Winning the Clean Energy Race: 2012* from the Pew Research Center. Online: http://www.pewenvironment.org/news-room/reports/whos-winning-the-clean-energy-race-2012-edition-85899468949.

45. See Cardwell, Diane. "For Solazyme, a Side Trip on the Way to Clean Fuel." *The New York Times*. 22 June 2013. Online: http://www.nytimes.com/2013/06/23/business/for-solazyme-a-side-trip-on-the-way-to-clean-fuel.html?_r=0. Also, note that much of this clean tech growth in recent years has been in the developing world: "According to David Wheeler at the Center for Global Development, developing countries now are responsible for two thirds of the new renewable energy capacity since 2002 in the world, and overall have more than half of the installed global renewable energy capacity." Gore, *The Future*, 346.

46. Tasch, Woody. Inquiries into the Nature of Slow Money: Investing as if Food, Farms, and Fertility Mattered. Chelsea Green Publishing, 2010.

47. Rifkin, *Third Industrial Revolution*, 120, 134–135.

48. Mason, *Meltdown*, 162–164.

49. See, for instance, Fossil Free Campaign: http://gofossilfree.org/.

50. Global Subsidies Initiative. "A How-To Guide: Measuring Subsidies to Fossil Fuel Producers," July 2010. Online: http://www.iisd.org/gsi/sites/default/files/pb7_ffs_measuring.pdf.

51. This term is increasingly used in sociology, ecological economics, and other domains to refer to economists wedded to deregulation and endless economic growth, despite the obvious negative consequences of this ideology.

BIBLIOGRAPHY

AASHE. https://stars.aashe.org/.

Anderson, Mark W. "Economics, Steady State." *Berkshire Encyclopedia of Sustainability*. Vol. 10, *The Future of Sustainability*. Eds. I. Spellerberg and D. Vasey. Great Barrington, MA: Berkshire Publishing, 2012.

Anielski, Mark. *The Economics of Happiness: Building Genuine Wealth*. Gabriola, BC: New Society, 2007.

Arndt, H. W. *The Rise and Fall of Economic Growth*. Melbourne: Longman Cheshire, 1978.

Bacon, Francis. "The Masculine Birth of Time." In *The Philosophy of Francis Bacon: An Essay on Its Development from 1603 to 1609, with New Translations of Fundamental Texts*. By Benjamin Farrington. Liverpool: Liverpool University Press, 1964. Originally published in 1605.

Baroni, L., L. Cenci, M. Tettamanti, and M. Berati. "Evaluating the Environmental Impact of Various Dietary Patterns Combined with Different Food Production Systems." *European Journal of Clinical Nutrition* 61 (2006): 279–286.

Bartlett, Albert A. "Forgotten Fundamentals of the Energy Crisis." *American Journal of Physics* 46 (September 1978): 876–888.

Bartlett, Albert A. "Reflections on Sustainability, Population Growth, and the Environment—Revisited." *Renewable Resources Journal* 15:4 (Winter 1997–1998): 6–23.

Beatley, Timothy. *Green Urbanism: Learning from European Cities*. Washington, DC: Island Press, 2000.

Beavis, Sara G. "Water." In *Berkshire Encyclopedia of Sustainability*. Vol. 10, *The Future of Sustainability*. Eds. I. Spellerberg and D. Vasey. Great Barrington, MA: Berkshire Publishing, 2012.

Bibliothèque Municipale de Bordeaux, 828 (C).

Bosselmann, Klaus, Daniel S. Fogel, and J. B. Ruhl, eds. *Berkshire Encyclopedia of Sustainability*. Vol. 6, *Measurements, Indicators, and Research Methods for Sustainability*. Great Barrington, MA: Berkshire Publishing, 2012.

Boulding, Kenneth. "The Economics of the Coming Spaceship Earth." In *Toward a Steady-State Economy*. Ed. Herman E. Daly. New York: Freeman, 1973. Originally published in Boulding, *Environmental Quality in a Growing Economy*. Ed. Henry Jarrett. Baltimore: Johns Hopkins University Press, 1966.

Boyd, David R. *Sustainability Within a Generation: A New Vision for Canada*. Vancouver: David Suzuki Foundation, 2004.

Braungart, Michael and William McDonough. *Cradle to Cradle: Remaking the Way We Make Things*. New York: North Point Press, 2002.

Brewer, Anthony. *The Making of the Classical Theory of Economic Growth*. London: Routledge, 2010.

Brown, Lester R. *Building a Sustainable Society*. New York: Norton, 1981.

Brown, Lester R. *Plan B 2.0: Rescuing a Planet under Stress and a Civilization in Trouble*. New York: Norton, 2006.

Brundtland, Gro Harlem. *Madam Prime Minister: A Life in Power and Politics*. New York: Farrar, Straus and Giroux, 2005.

Capital Institute. http://www.capitalinstitute.org/about.

Caradonna, Jeremy L. "Conservationism *avant la lettre*? Public Essay Competitions on Forestry and Deforestation in Eighteenth-Century France." In *Invaluable Trees: Cultures of Nature, 1660-1830*. Eds. Laura Auricchio, Heckendorn Elizabeth Cook, and Giulia Pacini. Oxford: SVEC, 2012, 39–54.

Caradonna, Jeremy L. *The Enlightenment in Practice: Academic Prize Contests and Intellectual Culture in France, 1670-1794*. Ithaca, NY: Cornell University Press, 2012.

Cardwell, Diane. "For Solazyme, a Side Trip on the Way to Clean Fuel." *New York Times*, 22 June 2013.

Carley, Michael and Ian Christie. *Managing Sustainable Development*. 2d ed. London: Earthscan, 2000.

Carlowitz, Hans Carl von. *Sylvicultura oeconomica, oder haußwirthliche Nachricht und Naturmäßige Anweisung zur wilden Baum-Zucht*. Leipzig: Braun, 1713.

Carrington, Damian. "Insecticide 'Unacceptable' Danger to Bees, Report Finds." *The Guardian*, January 16, 2013. Online: http://www.guardian.co.uk/environment/2013/jan/16/insecticide-unacceptable-danger-bees/print.

Carson, Rachel. *Silent Spring*. Boston: Houghton Mifflin, 1962.

Cavanag, J. and Mander, J. *Alternatives to Economic Globalization: A Better World is Possible*. 2d ed. San Francisco: Barret-Koehler, 2004.

City of Vancouver. "Greenest City: 2020 Action Plan." Online: http://vancouver.ca/files/cov/Greenest-city-action-plan.pdf.

Cleveland, Harlan, ed. *The Management of Sustainable Growth*. New York: Pergamon, 1981.

Columbia University, Graduate School of Arts and Sciences, Sustainable Development. http://gsas.columbia.edu/content/academic-programs/sustainable-development.

Commoner, Barry. *The Closing Circle: Nature, Man, and Technology*. New York: Knopf, 1971.

"Companies Increasingly 'Pursue Triple Bottom Line,'" *Environmental Leader*, May 1, 2013. Online: http://www.environmentalleader.com/2013/05/01/companies-increasingly-pursue-triple-bottom-line/.

Comte de Buffon, "Georges-Louis Leclerc." *Histoire naturelle, générale et particulière, avec la description du Cabinet du Roi*. 36 vols. Paris: Impr. Royale, 1749–1788.

Coomer, James C., ed. *Quest for a Sustainable Society*. New York: Pergamon, 1979.

Crist, Eileen. "Beyond the Climate Crisis: A Critique of Climate Change Discourse." *Telos* 141 (Winter 2007): 29–55.

Cronon, William. *Changes in the Land: Indians, Colonists, and the Ecology of New England*. New York: Hill & Wang, 1983.

Crosby, Alfred W. *Ecological Imperialism: The Biological Expansion of Europe, 900–1900*. Cambridge, UK: Cambridge University Press, 1986.

Daly, Herman E. *Steady-State Economics*. San Francisco, Freeman, 1977.

Daly, Herman E. "Toward Some Operational Principles of Sustainable Development." *Ecological Economics* 2 (1990): 1–6.

Daly, Herman E. "Farewell Speech to World Bank, 14 January 1994." Online: http://www.whirledbank.org/ourwords/daly.html.

Daly, Herman E. *Beyond Growth: The Economics of Sustainable Development*. Boston: Beacon Press, 1996.

Daly, Herman E. "Economics in a Full World." *Scientific American* 293.3 (2005): 100–107.

Daly, Herman E., ed. *Toward a Steady-State Economy*. New York: Freeman, 1973.

Darwin, Charles. *On the Origins of Species by Means of Natural Selection, or the Preservation of Favoured Races in the Struggle for Life*. London: J. Murray, 1859.

Darwin, Charles. *Descent of Man, and Selection in Relation to Sex*. London: J. Murray, 1871.

Descartes, René. *Discourse on Method and Meditations on First Philosophy*. 4th ed. Trans. Donald. A. Cress. Indianapolis: Hackett, 1998.

Diamond, Jared. *Guns, Germs, and Steel: The Fates of Human Societies*. New York: Norton, 1999.

Diamond, Jared. *Collapse: How Societies Choose to Fail or Succeed*. New York: Viking Press, 2005.

Diderot, Denis and Jean le Rond d' Alembert, eds. *Encyclopédie, ou Dictionnaire raisonné des sciences, des arts et des métiers*. 28 vols. Paris: Briasson, 1751–1772.

Dienst, Richard. *The Bonds of Debt*. New York: Verso, 2011.

Dillard, Jessie, Veronica Dujon, and Mary C. King, eds. *Understanding the Social Dimensions of Sustainability*. London: Routledge, 2008.

Downing, Louise. "Airlines Prepare to Take Off on Fuel Made from Algae, Wood Chips," July 6, 2011. Online: http://www.bloomberg.com/news/2011-2007-07/airlines-prepare-to-take-off-on-fuel-made-from-algae-wood-chips.html.

Dresner, Simon. *The Principles of Sustainability*. Earthscan, 2008.

Dryzek, John S. *Politics of the Earth: Environmental Discourses*, 2d ed. Oxford: Oxford University Press, 2005.

Earth Day Network. http://www.earthday.org/about.

Easterlin, Richard. "Does Economic Growth Improve the Human Lot? Some Empirical Evidence." In *Nations and Households in Economic Growth: Essays in Honor of Moses Abramowitz*. Eds. Paul A. David and Melvin W. Reder. New York: Academic Press, 1974.

EcoLogo. http://www.ecologo.org/en/.

Edelstein, Dan. *The Enlightenment: A Genealogy*. Chicago: University of Chicago Press, 2010.

Edwards, Andres R. *The Sustainability Revolution: Portrait of a Paradigm Shift*. Gabriola, BC: New Society Publishers, 2005.

Egan, Michael. *Barry Commoner and the Science of Survival: The Remaking of American Environmentalism*. Cambridge, MA: MIT, 2007.

Ehrenfeld, John R. *Sustainability by Design: A Subversive Strategy for Transforming our Culture*. New Haven, CT: Yale University Press, 2008.

Ehrlich, Paul R. *The Population Bomb*. New York: Buccaneer Book, 1995. Originally published in 1968.

Ehrlich, Paul R. and Holdren, John P. "Impact of Population Growth." *Science* n.s. 171.3977 (26 March 1971): 1212–1217.

Elkington, John. *Cannibals with Forks: The Triple Bottom Line of 21st Century Business*. Gabriola, BC: New Society, 1998.

Emerson, Ralph Waldo. "Self-Reliance." In *Essays: First Series*. Boston, MA: [n.p.], 1841.

Engels, Friedrich. *The Condition of the Working-Class in England 1844*. Trans. Florence Kelley Wischnewetzky. London: Allen and Unwin, 1968.

European Commission. Sustainable Development. http://ec.europa.eu/environment/eussd/escp_en.htm.

Evelyn, John. *Fumifugium, or, The Inconvenience of the Aer and Smoak of London Dissipated Together with some Remedies Humbly Proposed by J. E. esq. to His Sacred Majesty, and to the Parliament Now Assembled*. London: W. Godbid, 1661.

Evelyn, John. *Sylva, or a Discourse of Forest-Trees and the Propagation of Timber in His Majesty's Dominions*. 1st ed. London: Allestry and Martyn, 1664.

Fairtrade Foundation. "Aims of Fairtrade Standards." Online: http://www.fairtrade.net/aims-of-fairtrade-standards.html.

Fairtrade Foundation. "Fairtrade Bucks Economic Trend with 19% Sales Growth," February 25, 2013. Online: http://www.fairtrade.org.uk/press_office/press_releases_and_statements/february_2013/fairtrade_bucks_economic_trend.aspx.

Federal Ministry for the Environment, Nature Conservation and Nuclear Safety. *Development of Renewable Energy Sources in Germany 2011*, 2011–2012. Online: http://www.erneuerbare-energien.de/fileadmin/Daten_EE/Bilder_Startseite/Bilder_Datenservice/PDFs__XLS/20130110_EEiZIU_E_PPT_2011_FIN.pdf.

Federal Ministry for the Environment, Nature Conservation and Nuclear Safety. *Development of Renewable Energy Sources in Germany 2012*. Online: http://www.erneuerbare-energien.de/fileadmin/Daten_EE/Dokumente__PDFs_/ee_in_zahlen_ppt_en_bf.pdf.

Federal Ministry for the Environment, Nature Conservation and Nuclear Safety. *Renewable Energy Sources in Figures: National and International Development*, 2011. Online: http://www.erneuerbare-energien.de/fileadmin/ee-import/files/english/pdf/application/pdf/broschuere_ee_zahlen_en_bf.pdf.

Fiala, Nathan. "Measuring Sustainability: Why the Ecological Footprint Is Bad Economics and Bad Environmental Science." *Ecological Economics* 67.4 (2008): 519–525.

Flannery, Tim. *The Weather Makers: How We Are Changing the Climate and What It Means for Life on Earth*. Toronto: HarperCollins, 2006.

Florida, Richard. "Cities and the Creative Class." *City & Community* 2.1 (March 2003): 3–19.

Fodor, Eben. *Better Not Bigger*. Stony Creek, CT: New Society, 1999.

Foo, Kim Boon. "The Rio Declaration and Its Influence on International Environmental Law." *Singapore Journal of Legal Studies* 1992 (1992): 347–364.

Fossil Free Campaign. http://gofossilfree.org/.

Foucault, Michel. *L'archéologie du savoir*. Paris: Gallimard, 1969.

Fox, Stephen R. *John Muir and His Legacy: The American Conservation Movement*. Boston: Little, Brown, 1981.

Friedman, Milton. *Capitalism and Freedom*. Chicago: University of Chicago Press, 1962.

Galbraith, John Kenneth. *The Affluent Society*. Boston: Houghton Mifflin, 1958.

Gehl, Jan. *Cities for People*. Washington, DC: Island Press, 2010.

Georgescu-Roegen, Nicholas. *The Entropy Law and the Economic Process*. Cambridge, MA: Harvard University Press, 1971.

"German Solar Power Installations at Record High in 2012." Reuters, January 5, 2013. Online: http://www.reuters.com/article/2013/01/05/us-germany-solar-idUSBRE90406C20130105.

Global Footprint Network. http://www.footprintnetwork.org/en/index.php/GFN/.

Global Subsidies Initiative. "A How-To Guide: Measuring Subsidies to Fossil Fuel Producers." July 2010. Online: http://www.iisd.org/gsi/sites/default/files/pb7_ffs_measuring.pdf.

Goldblatt, David. *Sustainable Energy Consumption and Society: Personal, Technological, or Social Change?* Dordrecht: Springer, 2005.

Goodland, Robert, Herman E. Daly, Salah El Sarafy, and Bernd von Droste, eds. *Environmentally Sustainable Economic Development: Building on Brundtland.* Paris: UNESCO, 1991.

Gore, Al. *The Future: Six Drivers of Global Change.* New York: Random House, 2013.

Gottlieb, Robert. *Forcing the Spring: The Transformation of the American Environmental Movement.* Washington, DC: Island Press, 1993.

Green Analytics. http://greenanalytics.ca/index.php.

"Green Building Materials Will Reach $254 Billion in Annual Market Value by 2020, Forecasts Navigant Research," May 2, 2013. Online: http://www.marketwatch.com/story/green-building-materials-will-reach-254-billion-in-annual-market-value-by-2020-forecasts-navigant-research-2013-05-02.

Griggs, David, et al. "Sustainable Development Goals for People and Planet." *Nature.* 495 (21 March 2013): 305–307.

Grober, Ulrich. "Der Erfinder der Nachhaltigkeit." *Die Zeit* 48 (November 1999): 98. Online: http://www.zeit.de/1999/48/Der_Erfinder_der_Nachhaltigkeit.

Grober, Ulrich. *Deep Roots: A Conceptual History of "Sustainability."* Berlin: Wissenschaftszentrum Berlin für Sozialforschung, 2007.

Grober, Ulrich. *Sustainability: A Cultural History.* Trans. Ray Cunningham. Totnes, UK: Green Books, 2012.

Grove, Richard H. *Green Imperialism: Colonial Expansion, Tropical Island Edens and the Origins of Environmentalism, 1600–1860.* Cambridge, UK: Cambridge University Press, 1995.

Haeckel, Ernst. *Generelle Morphologie der Organismen.* Berlin: Reimer, 1866.

Hallsmith, Gwendolyn and Bernard Lietaer. *Creating Wealth: Growing Local Economies with Local Currencies.* Gabriola, BC: New Society, 2011.

Hansen, James, Makiko Sato Reto Ruedy Ken Lo, David W. Lea and Martin Medina-Elizade. "Global Temperature Change." *PNAS* 103.39 (26 September 2006): 14288–14293.

Happy Planet Index. *The (Un)Happy Planet Index.* Online: http://www.happyplanetindex.org/.

Hardin, Garrett. "The Tragedy of the Commons." *Science* n.s. 162.3859 (13 December 1968): 1243–1248.

Harris, Jonathan M., Timothy A. Wise, Kevin P. Gallagher, and Neva R. Goodwin, eds. *A Survey of Sustainable Development: Social and Economic Dimensions.* Washington, DC: Island Press, 2001.

Hawken, Paul. *The Ecology of Commerce: A Declaration of Sustainability.* New York: Harper Business, 1993.

Hawken, Paul. *Blessed Unrest: How the Largest Movement in the World Came into Being and Why No One Saw It Coming.* New York: Viking, 2007.

Hawken, Paul, Amory Lovins, and L. Hunter Lovins. *Natural Capitalism: Creating the Next Industrial Revolution.* Boston: Little, Brown, 1999.

Hayek, Friedrich. *The Road to Serfdom.* London: Routledge, 1944.

Heinberg, Richard. *The Party's Over: Oil, War and the Fate of Industrial Societies.* Gabriola, BC: New Society, 2003.

Heinberg, Richard. *Powerdown: Options and Actions for a Post-Carbon World.* Gabriola, BC: New Society, 2004.

Heinberg, Richard. *Peak Everything: Waking Up to the Century of Declines.* Gabriola, BC: New Society, 2007; reprint 2010.

Heinberg, Richard. *The End of Growth: Adapting to Our New Economic Reality.* Gabriola, BC: New Society, 2011.

Heinberg, Richard and Daniel Lerch, eds. *The Post-Carbon Reader: Managing the 21st Century's Sustainability Crises.* Healdsburg, CA: Watershed Media, 2010.

Hirsch, Fred. *Social Limits to Growth.* Cambridge, MA: Harvard University Press, 1976.

Hirsch, Robert L., Bezdek, Roger, and Wendling, Robert. "Peaking of World Oil Production: Impacts, Mitigation and Risk Management." Science Applications International Corporation/U.S. Department of Energy, February 2005. Online at: http://www.resilience.org/stories/2005-03-06/peaking-world-oil-production-impacts-mitigation-and-risk-management.

Holbach, Paul Henri Thiry, baron de. *Système de la nature, ou Des loix du monde physique et du monde moral.* London: M.-M. Rey, 1770.

Holling, C. S. "Resilience and Stability of Ecological Systems." *Annual Review of Ecology and Systematics* 4 (November 1973): 1–23.

Holmgren, David. *Permaculture: Principles and Pathways Beyond Sustainability.* Hepburn, Australia: Holmgren Design Services, 2002; reprint 2011.

Hopkins, Rob. *The Transition Companion: Making Your Community More Resilient in Uncertain Times.* White River Junction, VT: Chelsea Green Publishing, 2011.

"How Green Is It? How to Sort Out the Environmental Hype." *Globe and Mail,* 20 April 2012.

Hunt, Lynn. *Inventing Human Rights: A History.* New York: Norton, 2007.

International Energy Agency. *2012 Key World Energy Statistics.* Paris: International Energy Agency, 2012. Online: https://www.iea.org/publications/freepublications/publication/kwes.pdf.

Intergovernmental Panel on Climate Change. *Climate Change 2007: Synthesis Report.* Geneva: IPCC, 2007.

Intergovernmental Panel on Climate Change. *Renewable Energy Sources and Climate Change Mitigation*. Geneva: IPCC, 2011. Online: http://srren.ipcc-wg3.de/report.

Intergovernmental Panel on Climate Change. *The Synthesis Report of the Fifth Assessment Report*. 2014.

International Union for Conservation of Nature. *World Conservation Strategy: Living Resource Conservation for Sustainable Development*. Gland, Switzerland: IUCN, 1980.

Iwamoto, Junichi. "The Development of Japanese Forestry." In *Forestry and the Forest Industry in Japan*. Ed. Yoshiya Iwai. Vancouver: UBC Press, 2002.

Jackson, Tim. *Prosperity Without Growth: Economics For a Finite Planet*. London: Earthscan, 2009.

Jacobs, Jane. *The Death and Life of Great American Cities*. New York: Random House, 1961.

James, Sarah L. and Torbjorn Lahti. *The Natural Step for Communities: How Cities and Towns Can Change to Sustainable Practices*. Gabriola, BC: New Society, 2009.

Jevons, William Stanley. *The Coal Question: An Inquiry concerning the Progress of the Nation, and the Probable Exhaustion of our Coal-mines*. 2d ed. London: Macmillan, 1866. Originally published in 1865.

Johnson, Huey D. *Green Plans: Greenprint for Sustainability*. Lincoln: University of Nebraska Press, 2008.

Jones, Alison. "Selbstbestimmtes Leben: Hamburg's Rote Flora and the Roots of Autonomie in Twentieth-Century Germany." MA Thesis. University of Alberta, 2013.

Jones, Van. *The Green-Collar Economy: How One Solution Can Fix Our Two Biggest Problems*. New York: HarperOne, 2008.

Jowit, Juliette and Patrick Wintour. "Cost of Tackling Climate Change Has Doubled, Warns Stern." *The Guardian*, June 26, 2008. Online: http://www.guardian.co.uk/environment/2008/jun/26/climatechange.scienceofclimatechange.

Kagan, Sacha and Julia Hahn. "Creative Cities and (Un)Sustainability: From Creative Class to Sustainable Creative Cities." *Culture and Local Governance* 3.1–2 (2011): 11–27.

Keynes, John Maynard. *General Theory of Employment, Interest and Money*. New York: Macmillan, 1936.

Keynes, John Maynard. "Economic Possibilities for Our Grandchildren" (1930). In *John Maynard Keynes, Essays in Persuasion*. New York: Norton, 1963.

Hubbert, M. King. "Nuclear Energy and the Fossil Fuels." Presented at the Spring Meeting of the Southern District, American Petroleum Institute, Plaza Hotel, San Antonio, Texas, 7–9 March 1956.

Kiva. http://www.kiva.org/.

Kline, Benjamin. *First along the River: A Brief History of the U.S. Environmental Movement*. 4th ed. Lanham, MD: Rowman & Littlefield, 2011.

Klooster, Wim. *Revolutions in the Atlantic World: A Comparative History*. New York: New York University Press, 2009.

Kockerts, Kai. "The Fair Trade Story." Online: http://www.fairtrade.at/fileadmin/user_upload/PDFs/Fuer_Studierende/oikos_winner2_2005.pdf.

Kovacevic, V. and J. Wessler. "Cost-Effectiveness Analysis of Algae Energy Production in the EU." *Energy Policy* 38.10 (October 2010): 5749–5757.

Labrousse, Ernst. *Esquisse du mouvement des prix et des revenus en France au XVIIIe siècle*. Vol. 2. Paris: Librairie Dalloz, 1933.

Lang, Helen S. *Aristotle's Physics and its Medieval Varieties*. Albany: SUNY Press, 1992.

Laskawy, Tom. "Miracle Grow: Indian Farmers Smash Crop Yield Records Without GMOs." *Grist* February 22, 2013. Online: http://grist.org/food/miracle-grow-indian-farmers-smash-crop-yield-records-without-gmos/.

Layard, Richard. *Happiness: Lessons from a New Science*. New York: Penguin, 2006.

"LEED Buildings Grow by 14% Despite Market Crash." Greenbiz.com, November 17, 2010. Online: http://www.greenbiz.com/news/2010/11/17/leed-buildings-grow-14-percent-despite-market-crash.

Lefebvre, Henri. *Writings on Cities*. Eds. and trans. Eleonore Kofman and Elizabeth Lebas. Malden, MA: Blackwell, 1996.

Leopold, Aldo. *A Sand County Almanac: And Sketches Here and There*. Oxford: Oxford University Press, 1949.

Lewis, Michael and Pat Conaty. *The Resilience Imperative: Cooperative Transitions to a Steady State Economy*. Gabriola, BC: New Society, 2012.

Linnaeus, Carl. *Systema Naturae*. London: Natural History Museum, 1991. Originally published in multiple editions and volumes between the 1735 and the 1790s.

Lovejoy, T. E. and L. Hannah, eds. *Climate Change and Biodiversity*. New Haven, CT: Yale University Press, 2006.

Lovelock, James E. and Margulis, Lynn. "Atmospheric Homeostasis by and for the Biosphere: the Gaia Hypothesis." *Tellus* Series A, 26.1–2 (1 February 1974): 2–10.

Lovins, Amory. *Soft Energy Paths: Toward a Durable Peace*. New York: Penguin, 1977.

Lowe, Ian. "Shaping a Sustainable Future—An Outline of the Transition." *Civil Engineering and Environmental Systems*. 25.4 (December 2008): 247–254.

Lytle, Mark. *The Gentle Subversive: Rachel Carson, Silent Spring, and the Rise of the Environmental Movement*. Oxford: Oxford University Press, 2007.

MacKay, David J. C. *Sustainable Energy—Without Hot Air*. Cambridge, UK: UIT, 2008. Online: http://www.withouthotair.com/.

Malthus, Thomas Robert. *An Essay on the Principle of Population: Or a View of Its Past and Present Effects on Human Happiness; with an Inquiry into Our Prospects Respecting the Future Removal or Mitigation of the Evils which It Occasions*. 6th ed. London: J. Murray, 1798; reprint 1826.

Mandeville, Bernard. *The Fable of the Bees: or, Private Vices, Public Benefits*. London: Printed for Sam Ballard, at the Blue-Ball, in Little-Britain, 1705.

Mann, Charles C. *1493: Uncovering the New World Columbus Created*. New York: Knopf, 2011.

Marine Steward Council. http://www.msc.org/.

Marsh, George Perkins. *Man and Nature, or, Physical Geography as Modified by Human Action*. Ed. David Lowenthal. Seattle: University of Washington Press, 2003.

"Marsh, George Perkins." In *The Encyclopedia of the Earth*. Online: http://www.eoearth.org/view/article/154491/.

Marshall, Alfred. *Principles of Economics*. New York: Macmillan, 1890.

Mason, Paul. *Meltdown: The End of the Age of Greed*. New York: Verso, 2010.

Mckendrick, Neil, John Brewer, and J. H. Plumb, eds. *The Birth of a Consumer Society: The Commercialization of Eighteenth-Century England*. Bloomington: Indiana University Press, 1982; reprint 1985.

McKibben, Bill. "Buzzless Buzzword." *New York Times*, 10 April 1996.

McKibben, Bill. *Deep Economy: The Wealth of Communities and the Durable Future*. New York: Henry Holt, 2007.

Meadows, Dennis L. *Alternatives to Growth—I. A Search for Sustainable Futures*. Cambridge, MA: Ballinger, 1977.

Meadows, Donella H., Dennis L. Meadows, and Jørgen Randers (The Club of Rome). *The Limits to Growth: The 30-Year Update*. White River Junction, VT: Chelsea Green Publishing, 2004.

Meadows, Donella H., Dennis L. Meadows, Jørgen Randers, and William W. Behrens III (The Club of Rome). *The Limits to Growth*. New York: Universe Books, 1972.

Meadows, Donella H., Dennis L. Meadows, Jørgen Randers, and William W. Behrens III (The Club of Rome). *Limits to Growth: A Report for the Club of Rome's Project on the Predicament of Mankind*. 2d ed. New York: Universe Books, 1974.

Mill, John Stuart. *Principles of Political Economy*. London: John W. Parker, 1848.

Mill, John Stuart. *Principles of Political Economy with Some of Their Applications to Social Philosophy*. 7th ed. London: Longmans, Green, 1909.

Millennium Ecosystem Assessment. *Ecosystems and Human Well-Being: Biodiversity Synthesis*. Washington, DC: World Resources Institute, 2005.

Mishan, E. J. *The Cost of Economic Growth*. London: Staples, 1967.

Mishan, E. J. *Economic Growth Debate: An Assessment*. London: Allen & Unwin, 1977.

Mokyr, Joel. *The Enlightened Economy: An Economic History of Britain, 1700–1850.* New Haven, CT: Yale University Press, 2009.

Mollison, Bill and David Holmgren. *Permaculture One: A Perennial Agriculture for Human Settlements.* Tyalgum, Australia: Tagari, 1981.

Monbiot, George. "The Denial Industry." *The Guardian,* 19 September 2006.

Monbiot, George. *Heat: How to Stop the Planet from Burning.* New York: Random House, 2009.

Montrie, Chad. *A People's History of Environmentalism in the United States.* New York: Continuum, 2011.

Motavalli, Jim, Doug Moss, Brian C. Howard, and Karen Soucy, eds. *Green Living: The E Magazine Handbook for Living Lightly on the Earth.* London: A Plume Book, 2005.

Muir, John. *Our National Parks.* New York: Houghton Mifflin, 1901.

Muir, John. "The Hetch-Hetchy Valley." In *The Yosemite.* Chapter 12. New York: Century Company, 1912. Originally published in 1908 in the *Sierra Club Bulletin.*

Mumford, Lewis. *The City in History: Its Origins, Its Transformations, and Its Prospects.* New York: Harcourt, Brace & World, 1961.

Munby, Lionel M. *The Luddites and other Essays.* London: Micheal Katanka Books, 1971.

Nahi, Paul. "Government Subsidies: Silent Killer of Renewable Energy." *Forbes,* 14 February 2013.

Nash, Roderick. *The Rights of Nature: A History of Environmental Ethics.* Madison: University of Wisconsin Press, 1989.

National Resources Defense Council. "The Story of *Silent Spring,*" December 5, 2013. Online: http://www.nrdc.org/health/pesticides/hcarson.asp.

National Resources Defense Council. "Governments Should Phase Out Fossil Fuel Subsidies or Risk Lower Economic Growth, Delayed Investment in Clean Energy and Unnecessary Climate Change Pollution," June 2012. Online: http://endfossilfuelsubsidies.org/files/2012/05/fossilfuelsubsidies_report-nrdc.pdf.

Ocean Wise. http://www.oceanwise.ca/about.

Odum, Howard T. and Elisabeth C. Odum. *Energy Basis for Man and Nature.* New York: McGraw-Hill, 1976.

Ordonnance sur le fait des Eaux et Forêts. Paris: Chez P. Le Petit, 1669.

Oreskes, Naomi and Erik M. Conway. *Merchants of Doubt: How a Handful of Scientists Obscured the Truth on Issues from Tobacco Smoke to Global Warming.* New York: Bloomsbury, 2010.

Organic.org. "Certified Organic Label Guide." Online: http://www.organic.org/articles/showarticle/article-201.

Organic Trade Association. *Organic Trade Association's 2011 Organic Industry Survey.* See document summary online: http://www.ota.com/pics/documents/2011OrganicIndustrySurvey.pdf.

Ortiz, Isabel and Matthew Cummins. *Global Inequality: Beyond the Bottom Billion: A Rapid Review of Income Distribution in 141 Countries.* New York: UNICEF, 2011.

Owusu, K. and F. Ng'ambi. *Structural Damage: The Causes and Consequences of Malawi's Food Crisis.* London: World Development Movement, 2002.

Palley, Thomas. "Milton Friedman: The Great Laissez-Faire Partisan." *Economic & Political Weekly* 41.49 (9 December 2006): 5041–5043.

Pearce, David, Anil Markandya, and Edward B. Barbier. *Blueprint 1: For a Green Economy.* London: Earthscan, 1989.

Peterson, Thomas C., William M. Connolley, and John Fleck. "The Myth of the 1970s Global Cooling Scientific Consensus." *Bulletin of the American Meterological Society.* 89 (2008): 1325–1337.

Pew Research Center. *Who's Winning the Clean Energy Race: 2012.* Online: http://www.pewenvironment.org/news-room/reports/whos-winning-the-clean-energy-race- 2012-edition-85899468949.

Pezzey, John. *Sustainable Development Concepts.* Washington, DC: World Bank, 1992.

Pigou, Nicolas. *The Economics of Welfare.* London: Macmillan, 1932.

Pirages, Dennis Clark, ed. *The Sustainable Society: Implications for Limited Growth.* New York: Praeger, 1977.

Polèse, Mario and Richard E. Stren, eds. *The Social Sustainability of Cities: Diversity and the Management of Change.* Toronto: University of Toronto Press, 2000.

Postell, Sandra. "Water: Adapting to a New Normal." In *The Post-Carbon Reader: Managing the 21st Century's Sustainability Crisis.* Eds. R. Heinberg and D. Lerch. Healdsburg, CA: Watershed Media, 2010.

Rauschning, Dietrich, Katja Wiesbrock, and Martin Lailach. *Key Resolutions of the United Nations General Assembly: 1946–1996.* Cambridge, UK: Cambridge University Press, 1997.

Redefining Progress. Genuine Progress Indicator. Online: http://rprogress.org/sustainability_indicators/genuine_progress_indicator.htm.

Rees, William E. "Ecological Footprints and Appropriated Carrying Capacity: What Urban Economics Leaves Out." *Environment and Urbanization* 4.2 (October 1992): 121–130.

Rees, William E. "Impeding Sustainability? The Ecological Footprint of Higher Education?" *Planning for Higher Education* 31: 3 (2003): 88–98.

Rees, William E. "Thinking 'Resilience.'" In *The Post-Carbon Reader: Managing the 21st Century's Sustainability Crisis.* Eds. R. Heinberg and D. Lerch. Healdsburg, CA: Watershed Media, 2010.

Rees, William E. "True Cost Economics." In *Berkshire Encyclopedia of Sustainability.* Vol. 2, *The Business of Sustainability.* Eds. Chris Laszlo, Karen Christensen, and Daniel Fogel. Great Barrington, MA: Berkshire Publishing, 2010.

Resilience.org. http://www.resilience.org.

Ricardo, David. "On the Principles of Political Economy and Taxation." In *The Works and Correspondence of David Ricardo*. 3rd ed. Ed. P. Sraffa. Vol. 1. Cambridge, UK: Cambridge University Press, 1951.

Richards, John F. *The Unending Frontier: An Environmental History of the Early Modern World*. Berkeley: University of California Press, 2006.

Richardson, David. "Involuntary Migration in the Early Modern World." In *The Cambridge World History of Slavery*, Vol. 3. Eds. Keith Bradley and Paul Cartledge. Cambridge, UK: Cambridge University Press, 2011.

Rifkin, Jeremy. *The Third Industrial Revolution: How Lateral Power is Transforming Energy, the Economy, and the World*. New York: Palgrave Macmillan, 2011.

Righter, Robert W. *The Battle over Hetch-Hetchy: America's Most Controversial Dam and the Birth of Modern Environmentalism*. Oxford: Oxford University Press, 2005.

Robèrt, Karl-Henrik. *The Natural Step Story: Seeding a Quiet Revolution*. Gabriola, BC: New Society, 2002.

Robinson, John A. "Squaring the Circle? Some Thoughts on the Idea of Sustainable Development." *Ecological Economics* 48.4 (2004): 369–384.

Rocky Mountain Institute. http://www.rmi.org/Walmartsfleetoperations.

Rousseau, Jean-Jacques. *Discours sur les sciences et les arts*. Geneva: Barillot, 1750.

Rousseau, Jean-Jacques. *Discours sur l'origine et les fondements de l'inégalité parmi les hommes*. 1754.

Rousseau, Jean-Jacques. *Émile, ou de l'éducation*. Leipzig, Germany: Weidmann & Reich, 1762.

Sachs, Aaron. *The Humboldt Current: Nineteenth-Century Exploration and the Roots of American Environmentalism*. New York: Viking, 2006.

Sachs, Jeffrey D. *Common Wealth: Economics for a Crowded Planet*. New York: Penguin, 2008.

Sale, Kirkpatrick. *Rebels Against the Future: The Luddites and Their War on the Industrial Revolution: Lessons for the Computer Age*. Reading, MA: Addison-Wesley, 1995.

Schawb, Jim. *Deeper Shades of Green: The Rise of Blue-Collar and Minority Environmentalism in America*. San Francisco: Sierra Club Books, 1994.

Schendler, Auden. *Getting Green Done: Hard Truths from the Front Lines of the Sustainability Revolution*. New York: PublicAffairs, 2009.

Schendler, Auden. *Getting Green Done: Hard Truths from the Front Lines of the Sustainability Revolution*. New York: PublicAffairs, 2010.

Schumacher, E. F. *Small Is Beautiful: Economics as if People Mattered*. New York: HarperCollins, 1973.

Senick, Jennifer. "Green Building Benefits: By the Numbers." Rutgers Center for Green Building, 2011. Online: http://rcgb.rutgers.edu/uploaded_documents/Green.pdf.

Shapin, Steven. *The Scientific Revolution*. Chicago: University of Chicago Press, 1996.

Shapiro, Mark. "Conning the Climate: Inside the Carbon-Trading Shell Game." *Harper's Magazinez*, February 2010: 31–39.

Shellenberger, Michael and Ted Nordhaus. "The Death of Environmentalism: Global Warming Politics in a Post-Environmental World." *Special Issue: Don't Fear The Reapers: On the Alleged Death of Environmentalism. Grist*, 13 January 2005. Online: http://grist.org/article/doe-reprint/.

Shellenberger, Michael and Ted Nordhaus. *Break Through: From the Death of Environmentalism to the Politics of Possibility*. Boston: Houghton Mifflin, 2007.

Shovlin, John. *The Political Economy of Virtue: Luxury, Patriotism, and the Origins of the French Revolution*. Ithaca, NY: Cornell University Press, 2006.

Silverman, Howard. "Sustainability: The S-Word." *People and Place: Perspectives*, 15 April 2009. Online: http://www.peopleandplace.net/perspectives/2009/4/15/sustainability_the_s-word.

Skidelsky, Robert and Edward Skidelsky. *How Much Is Enough? Money and the Good Life*. New York: Other Press, 2012.

Smith, Adam. *An Inquiry into the Nature and Causes of the Wealth of Nations*. London: Printed for W. Strahan; and T. Cadell, 1776.

Smith, Alisa and J. B. Mackinnon. *The 100-Mile Diet: A Year of Local Eating*. Toronto: Random House Canada, 2007.

Southern Energy Management. http://www.southern-energy.com.

State of Green (Government of Denmark and energy producers). http://www.stateofgreen.com/en/Intelligent-Energy.

Steiguer, J. E. de. *The Origins of Modern Environmental Thought*. Tucson: University of Arizona Press, 2006.

Stern, Nicholas. "Stern Review: The Economics of Climate Change. Executive Summary." London: HM Treasury, 2006.

Stiglitz, Joseph E., Amartya Sen, and Jean-Paul Fitoussi. *Report by the Commission on the Measurement of Economic Performance and Social Progress*, 2009. Online: http://www.neweconomics.org/.

Stivers, Robert L. *The Sustainable Society: Ethics and Economic Growth*. Philadelphia: Westminster, 1976.

Stoll, Mark. "Rachael Carson's *Silent Spring*, A Book That Changed the World." Environment and Society Portal, Rachel Carson Center for Environment and Society at the University of Munich. Online: http://www.environmentandsociety.org/exhibitions/silent-spring/silent-spring-international-best-seller.

Stoll, Steven. *U.S. Environmentalism Since 1945: A Brief History With Documents*. Boston: Bedford/St. Martin's, 2007.

Surface Transportation Policy Project. Factsheet on Transportation and Climate Change. Online: http://www.transact.org/library/factsheets/equity.asp.

Tainter, Joseph. *The Collapse of Complex Societies*. Cambridge, UK: Cambridge University Press, 1988; reprint 2003.

Talloires Declaration. http://www.ulsf.org/programs_talloires.html.

Tasch, Woody. Inquiries into the Nature of Slow Money: Investing as if Food, Farms, and Fertility Mattered. Chelsea Green Publishing, 2010.

TerraChoice. *The Sins of Greenwashing: Home and Family Edition*, 2010. Online: http://sinsofgreenwashing.org/index.html.

The Green Life Online. "Greenwashing 101." Online: http://site.thegreenlifeonline.org/greenwash101/.

The National Textile Center. "Annual Report: Strategic Sustainability and the Triple Bottom Line," November 2008. Online: http://www.ntcresearch.org/pdf-rpts/AnRp08/S06-AC01-A8.pdf.

The Natural Step. "The Four Systems Conditions." Online: http://www.thenaturalstep.org.

The Post-Carbon Institute. http://www.postcarbon.org/.

"The Truth About Recycling." *The Economist*. 7 June 2007. Online: http://www.economist.com/node/9249262.

Thomas, Keith. *Man and the Natural World: Changing Attitudes in England, 1500–1800*. London: Allen Lane, 1983.

Thoreau, Henry David. "A Winter Walk." *The Dial* 4 (October 1843): 221–226.

Thoreau, Henry David. *Walden; Or, Life in the Woods*. Boston: Ticknor and Fields, 1854.

Thoreau, Henry David. *The Maine Woods*. Boston: Ticknor and Fields, 1864.

Trainer, Ted. *Renewable Energy Cannot Sustain a Consumer Society*. Dordrecht: Springer, 2007.

Trainer, Ted. *The Transition: To a Sustainable and Just World*. Sydney: Envirobook, 2010.

Trainer, Ted. "Can Renewable Energy Sustain Consumer Societies? A Negative Case." Simplicity Institute Report 12e, 2012.

Turner, Chris. *The Geography of Hope: A Tour of the World We Need*. Toronto: Vintage Canada, 2007.

Udall, Stewart L. *The Quiet Crisis*. New York: Holt, Rinehart and Winston, 1963.

United Nations. Declaration of the United Nations Conference on the Human Environment. New York: United Nations, 1972.

United Nations. Agenda 21. New York: United Nations, 1992.

United Nations. Rio Declaration on Environment and Development. New York: United Nations, 1992.

United Nations. Kyoto Protocol to the United Nations Framework Convention on Climate Change. New York: United Nations, 1997.

United Nations. *State of the World's Cities Report 2008/2009: Harmonious Cities*. New York: United Nations, 2008.

United Nations. United Nations Millennium Development Goals. New York: United Nations. Online: http://www.un.org/millenniumgoals/aids.shtml.

United Nations Framework Convention on Climate Change. "Fact Sheet: The Need for Mitigation." Online: http://unfccc.int/files/press/backgrounders/application/pdf/press_factsh_mitigation.pdf.

United Nations World Commission on Environment and Development. *Report of the World Commission on Environment and Development: Our Common Future.* New York: United Nations, 1987.

University of Alberta, Office of Sustainability. http://sustainability.ualberta.ca/.

University of Wisconsin Madison Urban and Regional Planning Department. "Urban Agriculture." Online: http://urpl.wisc.edu/ecoplan/content/lit_urbanag.pdf.

US Green Building Council. http://www.usgbc.org/ and http://www.usgbc.org/leed.

US Department of Agriculture. *Local Food Systems: Concepts, Impacts, and Issues.* Washington, DC: USDA, 2010. Online: http://permanent.access.gpo.gov/lps125302/ERR97.pdf.

US Department of Energy. *Building Technologies Program.* Washington, DC: US Department of Energy, Energy Efficiency and Renewable Energy, 2008.

US Environmental Protection Agency. "Causes of Climate Change." Online: http://www.epa.gov/climatechange/science/causes.html.

Vardi, Liana. *The Physiocrats and the World of the Enlightenment.* New York: Cambridge University Press, 2012.

Veblen, Thorstein. *The Theory of the Leisure Class.* New York: Dover Publications, 1994.

Victor, Peter. *Managing Without Growth: Slower by Design, Not Disaster.* Cheltenham, UK: Edward Elgar Publishing, 2008.

Wackernagel, Mathis, et al. "Tracking the Ecological Over-Shoot of the Human Economy." *Proceedings of the National Academy of Sciences* 99.14 (9 July 2002): 9266–9271.

Wackernagel, Mathis and William E. Rees. "Perceptual and Structural Barriers to Investing in Natural Capital: Economics from an Ecological Footprint Perspective." *Ecological Economics* 20.1 (January 1997): 3–24.

Wackernagel, Mathis and William E. Rees. *Our Ecological Footprint: Reducing Human Impact on the Earth.* Gabriola, BC: New Society, 1996.

Walker, Brian, C. S. Holling, Stephen R. Carpenter, and Ann Kinzig. "Resilience, Adaptability and Transformability in Social-Ecological Systems." *Ecology and Society.* 9.2 (2004): article 5.

Weisman, Alan. *The World Without Us.* New York: Thomas Dunne Books, 2007.

White, Gilbert. *The Natural History of Selbourne.* London: White, 1789.

White, Jr., Lynn. "The Historical Roots of Our Ecological Crisis." *Science* n.s. 155.3767 (10 March 1967): 1203–1207.

White, Hayden. *The Content of the Form: Narrative Discourse and Historical Representation*. Baltimore: Johns Hopkins University Press, 1987.

Whited, Tamara L. *Forests and Peasant Politics in Modern France*. New Haven, CT: Yale University Press, 2000.

Williams, Michael. *Deforesting the Earth: From Prehistory to Global Crisis*. Chicago: University of Chicago Press, 2002.

Williams, Raymond. "Ideas of Nature." In *Problems in Materialism and Culture: Selected Essays*. London: Verso, 1980.

Winter, Carl K. and Sarah F. Davis. "Scientific Status Summary: Organic Foods." *Journal of Food Science* 71.9 (2006): R117–R124.

Woody, Todd. "The U.S. Military's Great Green Gamble Spurs Biofuel Start-ups." *Forbes*, September 24, 2012. Online: http://www.forbes.com/sites/toddwoody/2012/09/06/the-u-s-militarys-great-green-gamble-spurs-biofuel-startups/2/.

World Commission on Environment and Development. *Report of the World Commission on Environment and Development: Our Common Future*. New York: United Nations, 1987.

Worster, Donald. *Nature's Economy: A History of Ecological Ideas*. 2d ed. Cambridge, UK: Cambridge University Press, 1994.

Worster, Donald. *The Wealth of Nature: Environmental History and the Ecological Imagination*. Oxford: Oxford University Press, 1994.

Wrigley, E. A. "The Limits to Growth: Malthus and the Classical Economists." *Population and Development Review* 14 (1988): 30–48.

Zehner, Ozzie. *Green Illusions: The Dirty Secrets of Clean Energy and the Future of Environmentalism*. Lincoln: University of Nebraska Press, 2012.

Zolli, Andrew and Ann Marie Healy. *Resilience: Why Things Bounce Back*. New York: Free Press, 2012.

Zovanyi, Gabor. *The No-Growth Imperative: Creating Sustainable Communities under Ecological Limits to Growth*. New York: Routledge, 2013.

INDEX

100-Mile Diet, 18, 227

"Agenda 21" (Rio Earth Summit, 1992),
    149, 155–157, 178
Agricultural Revolution, 3
Alabama, 157
algae. *See under* biofuels
Amazonian rainforest, 40
American Cyanamid Company, 97
American Enterprise Institute, 250
Anasazi Native Americans, 23
ancient Greece, 23
Anderson, Ray C., 170
Anielski, Mark, 183, 208, 212
Anthropocene, 3
Aquatic Species Program, 133
Arcadianism, 31, 52, 68
Arc de triomphe du Carrousel
    (Paris), 109
Arctic National Wildlife Refuge, 86
Aristotle, 28
Association for the Advancement of
    Sustainability in Higher Education
    (AASHE), 185, 206
Atlantic Northwest Cod fishery, 243
Atlantic Revolutions, 26
Australia, 226, 230

backcasting (Robèrt), 167
Bacon, Francis, 29
The Bahamas, 212
Barbados, 44
Barbier, Edward B., 158
Bartlett, Albert A., 11, 152
Bastiat, Frédéric, 130
Beatley, Timothy, 201
Behrens III, William, 14, 115
Beijing (China), 22, 71
Belgium, 26, 88, 107
Bentham, Jeremy, 77
Bertin, Henri, 48
Better Not Bigger (Fodor), 238
Bhopal industrial disaster (1984), 71, 174,
    282n70
biodiesel, 165, 191, 205
biodiversity: as aspect of sustainability,
    11, 13, 20, 142, 186, 214, 231, 242,
    246–247; climate change and,
    246–247; organic agriculture and,
    160; reforestation and, 244;
    resilience and, 240
biofuels: for airplanes, 205; algae-based,
    133, 165, 191, 195, 205, 252; ethanol
    and, 165, 181, 191; increasing use of,
    194; sustainability of, 164, 193–194